Christian Niedrig

Electrochemical Performance and Stability of $Ba_{0.5}Sr_{0.5}Co_{0.8}Fe_{0.2}O_{3-\delta}$ for Oxygen Transport Membranes

Schriften des Instituts für Angewandte Materialien –
Werkstoffe der Elektrotechnik
Karlsruher Institut für Technologie

Band 28

Eine Übersicht aller bisher in dieser Schriftenreihe
erschienenen Bände finden Sie am Ende des Buchs.

Electrochemical Performance and Stability of $Ba_{0.5}Sr_{0.5}Co_{0.8}Fe_{0.2}O_{3-\delta}$ for Oxygen Transport Membranes

by
Christian Niedrig

KIT Scientific Publishing

Dissertation, Karlsruher Institut für Technologie
Fakultät für Elektrotechnik und Informationstechnik, 2015

Impressum

KIT Scientific Publishing

Karlsruher Institut für Technologie (KIT)
KIT Scientific Publishing
Straße am Forum 2
D-76131 Karlsruhe

KIT Scientific Publishing is a registered trademark of Karlsruhe Institute of Technology. Reprint using the book cover is not allowed.

www.ksp.kit.edu

Print on Demand 2015

ISSN 2365-8029
ISBN 978-3-7315-0437-5
DOI: 10.5445/KSP/1000049670

Electrochemical Performance and Stability of $Ba_{0.5}Sr_{0.5}Co_{0.8}Fe_{0.2}O_{3-\delta}$ for Oxygen Transport Membranes

Zur Erlangung des akademischen Grades eines

DOKTOR-INGENIEURS

von der Fakultät für

Elektrotechnik und Informationstechnik
des Karlsruher Instituts für Technologie (KIT)

genehmigte

DISSERTATION

von

Dipl.-Phys. Christian Niedrig
geb. in Nürnberg

Tag der mündlichen Prüfung: 13.05.2015
Hauptreferentin: Prof. Dr.-Ing. Ellen Ivers-Tiffée
Korreferentin: Prof. Dr. Dagmar Gerthsen

To my parents

Acknowledgements

This thesis developed and grew at the Institute for Applied Materials – Electrical and Electronic Engineering (IAM-WET), formerly Institute of Materials for Electrical and Electronic Engineering (IWE) at the Karlsruhe Institute of Technology (KIT). My deepest heartfelt gratitude belongs to Prof. Dr.-Ing. Ellen Ivers-Tiffée, who went above and beyond just the truly excellent scientific working conditions provided at her institute in her unwavering trust and personal support at all times during my work on this thesis.

My appreciation and big thanks also go to Prof. Dr. Dagmar Gerthsen, Laboratory for Electron Microscopy (LEM) at KIT, for her agreement to serve as second reviewer for this work and the fruitful cooperation and frequent discussions with her and her group during several research projects.

I would also like to thank Prof. Bernhard Holzapfel (chair), Prof. Wolfgang Heering, and Prof. Manfred Thumm (all KIT) for their agreement to serve on the board of examiners.

My great appreciation and thanks are owed to my group leader Dr.-Ing. Stefan F. Wagner, who has always acted as sounding board for crazy ideas and all kinds of problems. Discussions with him nearly always resulted in great solutions, new perspectives on retrieved data and were always led in friendly banter while never losing focus on the scientific work. Often, these discussions were joined by Dr.-Ing. Wolfgang Menesklou, who perfectly complemented our round by introducing even more viewpoints and always lightening the mood with his easygoing and sunny attitude. Working closely with the both of them has been and still is a great privilege, for which I am truly grateful.

I also want to thank Dr. Monica Burriel, currently at Laboratoire des Matériaux et du Génie Physique (LMGP), Grenoble Institute of Technology, Grenoble, France, and Dr. Koichi Asano, Central Research Institute of Electric Power Industry (CRIEPI), Tokyo, Japan, for their shared experience and great cooperation during their stay at our institute. This work profited a lot from their expertise and their friendly nature. I look forward to meeting them again at various international conferences.

At LEM, I especially want to thank Dr.-Ing. Heike Störmer, Dr. Philipp Müller, and Matthias Meffert for their past and ongoing cooperation with me and our group in several research projects. Numerous discussions, joint publications and mutual conference attendances brought new insights into the investigated material systems.

For their shared experience and uncomplicated supplying with samples as well as performed measurements on our samples and their pleasant and friendly attitude during lots of project- and also bilateral meetings, my thanks go to Dr. Stefan Baumann and Dr. Wilhelm Meulenberg, both from the Institute of Energy and Climate Research (IEK-1) at Forschungszentrum Jülich (FZJ), Jülich. Planning new measurements with them has always been a pleasure. Also from FZJ, in this case during his stay at the Ernst Ruska-Centre (ER-C), Dr. Stefan Roitsch has been especially helpful by providing great electron microscopy images in the context of several research projects.

Lots of knowledge, pointers, and suggestions, as well as fruitful meetings during projects and conferences, and also joint measurements are reason for lots of thanks toward Prof. H.J.M. Bouwmeester and his coworkers at the University of Twente, Netherlands.

This work would not have been possible without the competent work of numerous student assistants over the years and especially the diligent commitment of the several students I had the honor of supervising during their Study-, Bachelor- or Diploma theses: Simon Taufall, Philipp Braun, Isabelle Hofstetter, Lana Unger and Lennart Willms all deserve special mention and thanks for their great work and professional attitude. I am especially glad that P. Braun and L. Unger decided to stay at our institute, researching for their own PhD thesis. Working with them has always been a joyful activity.

The backbone of the institute in the form of the mechanical workshop, electronic workshop, chemical laboratory, and secretariat deserves to be mentioned. The tireless and flawless work of Stefan Ziegler and his coworkers (mechanical), Christian Gabi and Torsten Johannsen (electronic) and Sylvia Schöllhammer, followed by Sarah van den Hazel (chemical), and Andrea and Marika Schäfer (secretariat) played a big part in the realization of this thesis. My thanks go to all of them.

All former and present colleagues at the IAM-WET deserve my thankfulness for their friendliness and cooperative attitude, making my stay at the institute a very pleasant experience.

Last but not least, my deepest appreciation and profound gratitude goes to my parents, who always encouraged me to follow my dreams and never wavered in their unconditional support along the way. Thank you!

Vobis gratias ago!

Karlsruhe, July 2015 Christian Niedrig

Table of Contents

Chapter 1

Introduction

1.1 Anthropogenic Greenhouse Gas Emissions

Ever since the industrial revolution, anthropogenic greenhouse gas emissions into the atmosphere keep increasing. While the ongoing climate change is mainly driven by emissions of CO_2 (estimated to be 60% of the total due to all anthropogenic greenhouse gases [1]), some other greenhouse gases also play a part. After CO_2, methane is the next most important anthropogenic greenhouse gas, accounting for 20% of the total impact as of 2001 [1]. One important source of methane is the energy sector, but agriculture and domestic waste disposal contribute more to the global total [2]. Also nitrous oxide contributes directly to climate change albeit only amounting to 6% of the total impact of all greenhouse gases. The main source is agriculture but also industrial production of select chemicals.

While some of the anthropogenic CO_2 stems from industrial processes and some from land use changes (mainly deforestation), the main contribution comes from the combustion of all kinds of fossil fuels, with the emissions continuing an upward trend in the early years of the 21st century.

Fossil fuels are the dominant form of energy utilized in the world (86%), therefore accounting for about 75% of global anthropogenic CO_2 emissions in 2001 [2]. In 2002, 149 Exajoules (EJ[1]) of oil, 91 EJ of natural gas, and 101 EJ of coal were consumed worldwide [3], with global primary energy consumption steadily growing at an average rate of 1.6% annually between 1995 and 2001 [4]. This resulted in an increase of average global CO_2 emissions by 1.4% per year in the same time period, with the total emissions from fossil fuel consumption

[1] 1 EJ = 10^{18} J

and flaring of natural gas amounting to 24 Gt of CO_2 in the year 2001, with industrialized countries being responsible for 47% of all energy-related CO_2 emissions [4].

Estimations of the International Energy Agency (IEA) in 2003 as to the CO_2 emissions from various sources worldwide show power generation to be the single largest source of emissions. Another sector where emissions arise from a few large point sources is in parts of Manufacturing and Construction [4].

The second largest source of CO_2 emissions is transport, growing even faster than those emissions from energy and industry in the last few decades [5]. The main difference, however, is the fact that transport emissions stem mainly from a large amount of small, distributed sources. This renders the transport emissions not feasible to be a candidate for the possible uses of CO_2 capture and storage technology that will be introduced in the next part of this chapter.

1.2 CO_2 Capture and Storage

1.2.1 Overview

While CO_2 emissions can be reduced by more efficient use of fossil fuel or different technologies for the production of energy like nuclear power or from renewable sources (e.g. water and wind power, solar and geothermal energy, biomass consumption etc.), the fact remains that in the near and middle future fossil fuel will globally still remain the most used energy source. Therefore the idea was born to try and separate CO_2 from industrial and energy-related sources and transport the purified carbon dioxide to a storage location where a long-term isolation from the atmosphere would be guaranteed.

In 2005, the Intergovernmental Panel on Climate Change (IPCC) published a special report on carbon dioxide capture and storage (CCS) [6]. Here, it could be seen that the largest amounts of CO_2 emitted through the consumption of fossil fuels stem from relatively few large point sources, that is approx. 7000 stationary sources world-wide from power generation, cement production, and oil refining being responsible for a total emission of more than 12 Gt of CO_2 per year. These large stationary sources would be the ideal candidates for a possible capture of CO_2.

1.2.2 Storage Options

Several storage options can be discussed, the first being the injection of the pure CO_2 into the oceans. Carbon dioxide is soluble in water, therefore the oceans steadily take up CO_2 from the atmosphere (approx. 7 Gt per year in the last decade), amounting to 500 Gt of anthropogenic emissions during the last 200 years. The naturally absorbed CO_2 already accounts for a drop in the pH value of 0.1 in the upper oceans. Possible injections could be done by ship or pipeline in depths above approx. 3 km – where the CO_2 would dissolve in the water – or below that depth, where the water pressure would be high enough to liquefy CO_2, therefore forming CO_2 lakes at the seabed. However, serious impacts on ocean chemistry and marine life are to be expected, as large-scale injections would decrease the pH significantly at the injection site, with expected immediate mortality of marine life. Furthermore, no ecosystem experiments have yet been conducted in the deep ocean. Also, the long term sequestration of the carbon dioxide from the atmosphere is not given using this storage option, as even the liquid CO_2 will eventually dissolve in the sea water and in the long term equilibrate with the atmosphere. Projections show that even with an injection depth of 3 km, after 500 years only 71% of the injected CO_2 will still be contained [6].

A probably more feasible option is storage in geological formations. While CO_2 can be – and already is – used in enhanced oil and gas recovery (injecting CO_2 into the reservoir presses out the fossil fuel) and enhanced coal bed methane recovery (ECBM, where the injected CO_2 adsorbs onto coal replacing gases like methane), sequestration can also be achieved by using depleted oil and gas reservoirs as storage space or even by injecting the carbon dioxide into deep saline aquifers both onshore and offshore. Fig. 1.1 shows an overview of these possible techniques.

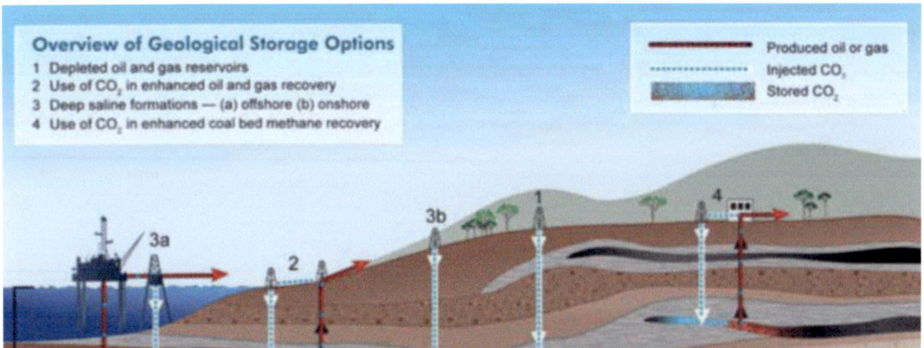

Fig. 1.1: Overview of geological storage options for separated carbon dioxide [6].

In the two cases of using depleted reservoirs and deep saline formations, injection normally takes place at depths below 800 meters where CO_2 will be in the liquid state with a density range from 50 to 80% of the density of water. Buoyant forces necessitate the presence of a dense cap rock that physically traps the carbon dioxide. However, capillary forces can retain the CO_2 in pore spaces and a fraction forms solid carbonates with rock minerals. The carbon dioxide will also slowly dissolve in the in-situ water over hundreds to thousands of years, with the CO_2-rich water sinking down and further trapping the gas. Industrial scale projects testing these options are already ongoing, e.g. the In Salah project in Algeria, the Sleipner project in Norway or the Gorgon project in Australia [6]. Fig. 1.2 shows possible storage locations in Western Europe, among those oil and gas fields, unminable coal seams suitable for ECBM, and deep saline formations with potential storage capacities of up to 900 Gt, 200 Gt, and 10000 Gt carbon dioxide, respectively.

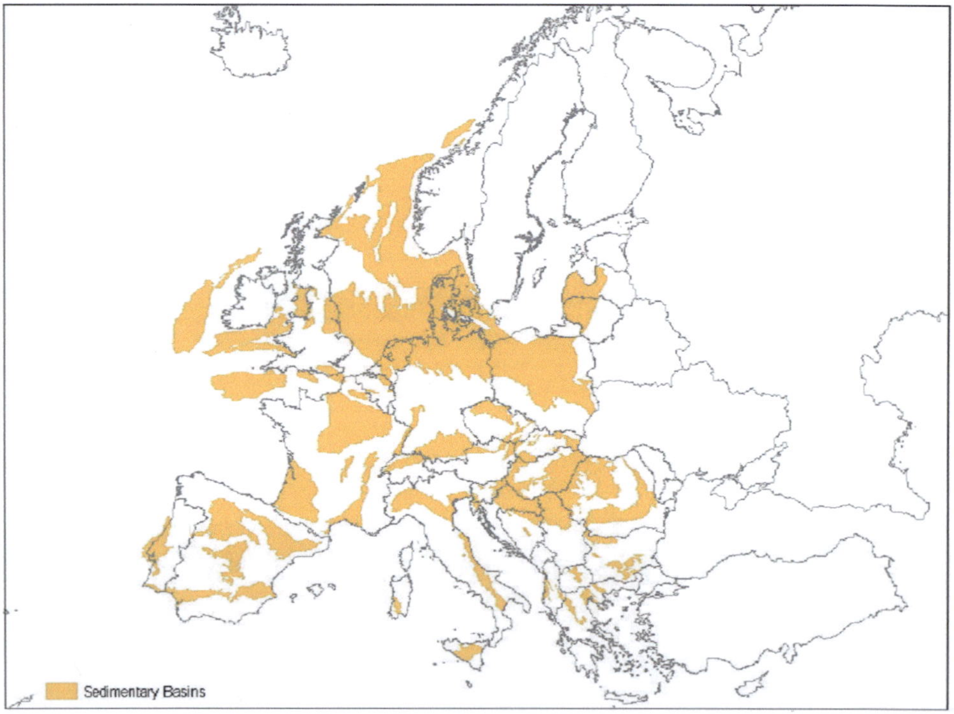

Fig. 1.2: Sedimentary basins suitable for CO_2 storage in Western Europe [6].

However, using these possible storage locations remains to be regulated by local governments, which have to address a multitude of different topics from safety to local environmental concerns and concerns in the population.

As available storage space is somewhat limited and should be used as efficient as possible, it is prudent to obtain CO_2 from a combustion process in the highest possible purity. This is even more important considering the transport costs to the injection site. Estimates show them to be approx. 6 US$/t for pipeline transport (offshore/onshore, depending on terrain conditions) and double that value for ship transport [6].

1.2.3 Obtaining Pure CO_2 from a Combustion Process

As the importance of cost and storage efficiency dictates, CO_2 has to be obtained in a very high purity. The first possibility that comes to mind is post combustion separation of the flue gases emitted from a power plant. This is an interesting concept, as the plant itself would not have to be severely changed but could simply be retrofitted with a further process of flue gas treatment: as soot and SO_X are already filtered from the flue gas in modern power plants, just another step for water vapor condensation and $N_2/NO_X/CO_2$ separation would have to be added. Also, a chemical absorption process for CO_2 separation from other gases already exists and is used on a large scale for separating carbon dioxide from natural gas directly after extraction. Therefore, using chemical absorption also plays a dominant role in current R&D activities on CCS [6]. Monoethanolamine (MEA) is used to absorb up to 90% CO_2 from the flue gas in an absorber at 40-70 °C, then heated up in a stripper unit to 100-120 °C in order to release the CO_2. After cooling, most of the MEA can be reused, slight losses occur by reaction with present traces of SO_2. Thus, several weaknesses present themselves in this method. The solvent regeneration process presents high energy demands, while degeneration of the solvent introduces high material and disposal costs. Furthermore, the used solvents create additional environmental pollution.

Calculations have been made for the MEA process used in a 1100 MW coal fired power plant, where the flue gas flow of 3 million Nm^3/h contains 800 t/h CO_2. This would require an initial fill of MEA of approx. 1800 t with a needed refill rate of at least 1.3 t/h and the additional energy demand would result in estimated efficiency losses $\Delta\eta$ for the overall power plant efficiency of around 12-14 percentage points [7]. Other simulations for a 600 MW coal fired power plant yielded efficiency losses of 8.2 percentage points [8].

Alternatively, gas separation membranes can be used for post-combustion CO_2 separation, possessing the advantages of end-of-pipe application and reduced environmental impact compared to the chemical absorption method [9;10]. Gas separation membranes for use in post-combustion capture have been investigated independently by several groups [11-17]. Simulations for a cascade membrane concept in the abovementioned 600 MW reference plant showed the efficiency loss to be only 6.4 percentage points [8].

There are, however, other possible means to obtain pure CO_2 from a combustion process. Fig. 1.3 shows an overview over several techniques. In the pre combustion process, the fuel is treated by steam reformation and water-gas shift reaction to form a mixture of H_2 and CO_2, which would have to be separated in order to obtain pure CO_2. The hydrogen is then used for heat and power generation. Different methods for this type of separation process are possible, with physical absorption processes yielding efficiency drops of 5-11 percentage points [8], while other possibilities are under investigation by researchers around the world, e.g. the development of zeolite membranes for nano-sieving [18-22] or high temperature proton conductors [23-25].

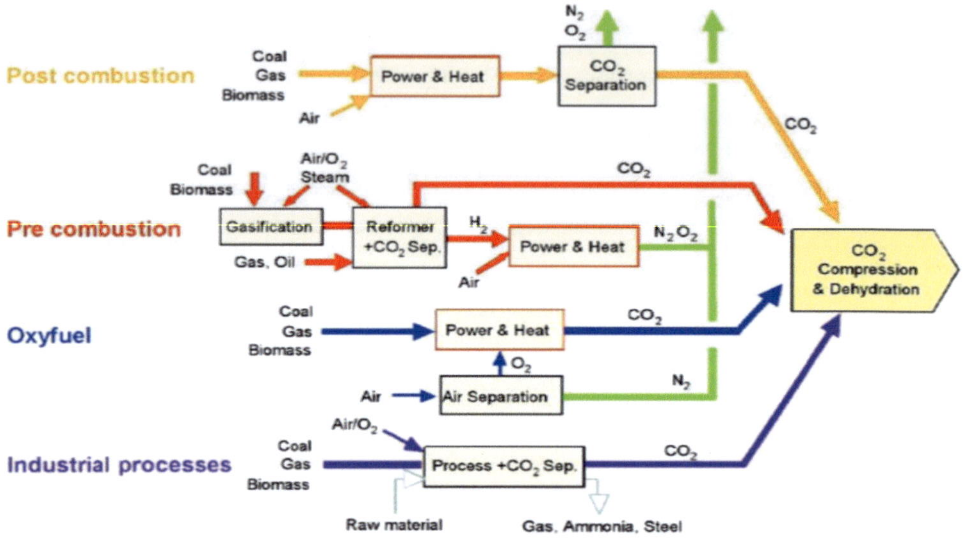

Fig. 1.3: Overview of different processes to gather pure CO_2 from fossil fuel combustion by gas separation [6].

In the Oxyfuel process, a different concept is used, owing to the fact that the main portion of the flue gas of a fossil fuel power plant consists of nitrogen and NO_X. This whole part could be avoided by not using air to enable combustion, but burning the fuel in pure oxygen. That way, the resulting flue gas would mainly contain CO_2, soot, water vapor, and SO_X, the latter three of which are either already filtered out by the present flue gas treatment in modern power plants (soot, SO_X) or could easily be done so by condensation (in case of the water vapor), leaving nearly pure CO_2 for subsequent transportation and storage.

For this Oxyfuel process, the gas separation challenge comes down to the production of pure oxygen from air. While cryogenic air separation is a well-established process for this task, the energy demands for use in a power plant would amount to an efficiency loss of 8-11 percentage points [26]. There is, however, the possibility to use oxygen transport membranes (OTM) for the air separation process in order to obtain the necessary oxygen, reducing the efficiency loss by 2-5 percentage points [8] in comparison – for example, Stadler et al. report achievable efficiencies of 40.1% and 40.7% for a power plant using OTMs in comparison to 45.9% for the reference plant without CO_2 sequestration [27]. Using the Oxyfuel process will, however, require major modifications to the base power plant, i.e. due to generally higher burning temperatures and the integration of high temperature membranes in the latter case.

1.3 Membranes for Oxygen Separation

1.3.1 Mixed Ionic Electronic Conducting (MIEC) Oxides

Most promising materials for oxygen transport membranes are mixed ionic electronic conducting perovskites ABO_3 (with A: Sr, Ba, La, … and B: Fe, Co, …) which show a remarkable flexibility and stability with respect to cation substitution (thus enabling custom-tailoring of material properties for different applications) and are able to excellently conduct oxygen ions via numerous vacancies in the anionic sub-lattice and compensate for the charge transfer by fast simultaneous electronic transport, which has been proven long since (e.g., [28]). These transports occur continuously, as long as a gradient in the chemical oxygen potential (in this case oxygen partial pressure pO_2) across the gas tight material is applied (typical values for the technical application in a three end module – cf. next section – would be $1\text{ bar} \leq pO_2^{(1)} \leq 5\text{ bar}$, $pO_2^{(2)} \approx 10^{-3}$ bar). This is economically feasible only if the oxygen in- and excorporation at the membrane surface is governed by a sufficiently high chemical oxygen surface exchange coefficient $k^{\delta}(pO_2,T)$ and if fast ionic oxygen transport through the lattice is guaranteed by a high chemical oxygen diffusion coefficient $D^{\delta}(pO_2,T)$. Fig. 1.4 shows a schematic of such a mixed conducting membrane.

These perovskite-type oxides hold promise for various high-temperature applications, such as solid oxide fuel cell (SOFC) electrodes (e.g., [29-31]), resistive-type gas sensors, (e.g., [32;33]) as well as oxygen transport membranes, (e.g., [34;35]). In all of these applications the material should exhibit excellent oxygen transport properties while being stable under the ambient operating conditions of temperature (typically in the range between 600…900 °C) and gas atmosphere (mainly low oxygen partial pressure, but sometimes also in various other gases), and being compatible with other materials with which it is in direct contact – often a non-trivial task in view of the relatively high operation temperatures.

Oxygen transport through such a MIEC membrane of the thickness L is usually given by the Wagner equation:

$$j_{O_2} = -\frac{RT}{4^2 F^2 L} \int_{\ln pO_2^{(1)}}^{\ln pO_2^{(2)}} \frac{\sigma_{el}\sigma_{ion}}{\sigma_{el}+\sigma_{ion}} d\ln pO_2 \tag{1.1}$$

where R is the ideal gas constant, F is the Faraday constant, $pO_2^{(1)}$ and $pO_2^{(2)}$ are the oxygen partial pressures on feed and permeate side of the membrane and σ_{el} and σ_{ion} are the electronic and ionic conductivity, respectively [34].

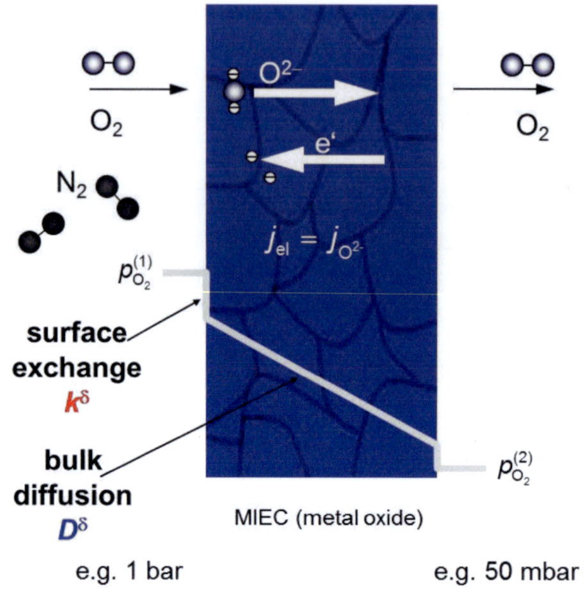

Fig. 1.4: Schematic of a mixed ionic-electronic conducting oxygen transport membrane. Given are examples for the absolute pressures on both sides of the membrane, for example as used in a four-end module (cf. next section).

The surface exchange at the gas/membrane interface on both sides of the membranes can be described using the pO_2-dependent surface exchange coefficient $k^{\delta}(pO_2,T)$:

$$\vec{n}\cdot\vec{j}_{\text{diff,O}^{2-}}(\vec{x}) = -k^{\delta}(pO_2,T)\cdot(c_{\text{O}^{2-},\text{eq}}(\vec{x},t) - c_{\text{O}^{2-}}(\vec{x})) \tag{1.2}$$

where $\vec{n}$ is the unit vector normal to the gas/MIEC interface and $c_{\text{O}^{2-},\text{eq}}$ the equilibrium oxygen ion concentration that can be calculated from oxygen nonstoichiometry measurements as described in our recent paper [36].

If the electronic conductivity in the MIEC material is high enough so that local electroneutrality can be assumed, the ionic diffusion in the membrane volume is driven by a concentration gradient in the volume itself and can be described by Fick's law using the chemical diffusion coefficient $D^{\delta}(pO_2,T)$:

$$\vec{j}_{\text{diff,O}^{2-}}(\vec{x}) = -D^{\delta}(pO_2,T)\cdot\nabla c_{\text{O}^{2-}}(\vec{x}) \tag{1.3}$$

Therefore, if the material parameters $k^{\delta}(pO_2,T)$ and $D^{\delta}(pO_2,T)$ are known for the relevant oxygen partial pressures and temperatures and the equilibrium oxygen ion concentration can be derived, the total oxygen flux through a membrane can then be calculated by integration of the ionic current density over the surface cross-sectional area A – even in case of a porous membrane by also considering gas diffusion in the pores via, e.g., the dusty gas model as shown in [36]:

$$j_{O_2}(pO_2,T) = \frac{1}{2}\int_A \vec{n}\cdot\vec{j}_{\mathrm{diff,O^{2-}}}(\vec{x})dA \tag{1.4}$$

Transport parameters as well as oxygen stoichiometry changes in the MIEC material are therefore important information to evaluate the performance of a possible oxygen transport membrane. Parts of this work contribute to this quest.

1.3.2 Membrane Module Concepts

As has been shown, the oxygen transport through the membrane has to be facilitated by applying a gradient in oxygen partial pressure to the MIEC material. For an OTM integrated into a combustion process, this can be done in two basically different ways [8;27] as shown in Fig. 1.5. While in both the depicted four-end module and the three-end module the feed side of the membrane is flushed with compressed ambient air and the separated oxygen is then used to enrich the oxygen concentration in the flue gas and use it again to deliver the oxidant to the firing process (therefore steadily increasing CO_2 concentration for subsequent storage), in the first case the permeate side has to be in contact with the oxygen deprived flue gas from the combustion process at ambient air pressure. The gradient in oxygen partial pressure is then due to the combination of absolute pressure difference as well as concentration difference due to the oxygen depleted exhaust gas. The membrane does not have to be artificially heated to the intended operating temperature, as the hot flue gas takes care of that necessity. However, direct contact with components of the exhaust like soot, SO_X or even the highly concentrated carbon dioxide may be detrimental to the membrane material if chemical stability is not guaranteed under these conditions.

The three-end module concept is fundamentally different in that it is basically a stand-alone unit that produces pure oxygen – that then can but does not have to be used to enrich the oxygen content in the flue gas. Pure oxygen can also be used to directly burn fuel or for completely different applications. The partial pressure gradient is achieved by applying a vacuum to the permeate side of the membrane in using a vacuum pump, in that way avoiding any contact to potentially detrimental chemical compounds. This however comes with the cost of requiring artificial heating for the membrane and feed gas and additional efficiency losses due to the vacuum pump system.

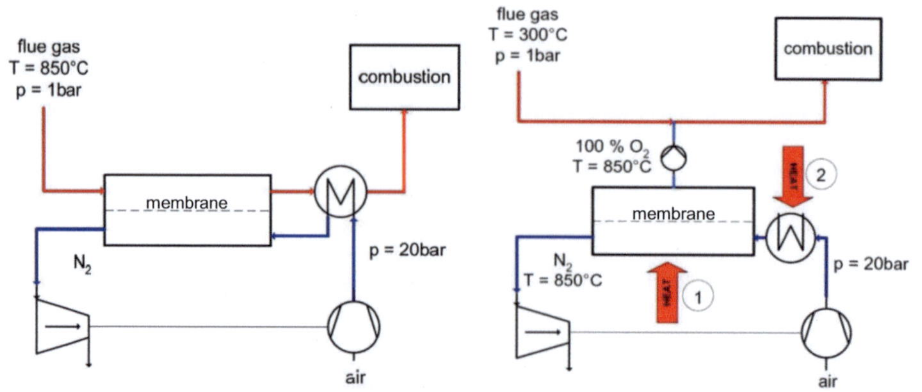

Fig. 1.5: Pressure driven four-end (left) and three-end (dead-end) module concept (right) for the possible usage of gas separation membranes [37].

1.3.3 MIEC Material Selection

The cubic modification of the $SrCo_{1-x}Fe_xO_{3-\delta}$ was firstly identified as highly promising in 1985 [28] but was found to undergo a phase transition to an orthorhombic Brownmillerite structure due to oxygen vacancy ordering [38;39]. This reduces the oxygen ion mobility and there have since been numerous attempts to stabilize the cubic phase by A-site doping. Barium turned out to be a useful element as it also enhances the ionic oxygen conductivity [40]. Due to its larger ionic radius, however, the introduced deformation of the perovskite structure favors the formation of hexagonal barium oxide in the case of too high doping levels. Therefore, a compromise has to be made between high transport capabilities in the case of high barium concentration and phase stability with low dopant content.

This led to the state-of-the-art material system of $Ba_{0.5}Sr_{0.5}Co_{0.8}Fe_{0.2}O_{3-\delta}$ (BSCF), which possesses formerly unobtainable oxygen transport capabilities [41] in its cubic phase (e.g. [40-50]),while being chemically stable even at very low oxygen partial pressures (cf. Chapter 3 and [51]). Other widely used MIEC perovskites are also well known from SOFC research, such as $La_{0.58}Sr_{0.4}Co_{0.2}Fe_{0.8}O_{3-\delta}$ (LSCF) (e.g., [52-57]), $Pr_{0.58}Sr_{0.4}Co_{0.2}Fe_{0.8}O_{3-\delta}$ (PSCF) [58;59], or $La_{0.6}Sr_{0.4}CoO_{3-\delta}$ (LSC) (e.g., [31;60-64]), with LSCF being the material of choice for state-of-the-art cathodes, PSCF having been considered for use as alternative intermediate-temperature cathodes, and LSC recently showing the highest performance as a thin-film SOFC cathode with extraordinary electrochemical properties [31].

For use as a material for oxygen transport membranes, however, great interest has accrued to BSCF due to its high oxygen permeation properties [41]. In Fig. 1.6 it can be seen that a BSCF bulk membrane with a thickness of 1 mm already outperforms the oxygen flux obtainable with a

20 µm thick thin film made from LSCF. With thinner BSCF bulks and finally supported BSCF thin films, the figure of merit set during the Helmholtz project MEM-BRAIN (cf. next section) of an oxygen flux of 10 ml_N min^{-1} cm^{-2} can finally be obtained. Even higher fluxes are possible by the application of an additional activation layer as described in [41]. The main focus of this work has therefore been on the BSCF system.

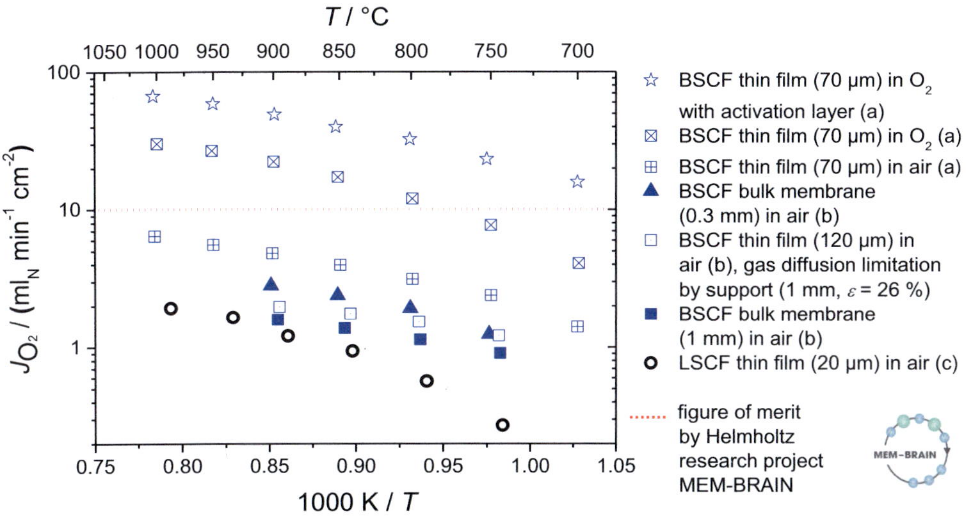

Fig. 1.6: Oxygen fluxes through BSCF bulk and thin film membranes ((a) [41], (b) [65]) in comparison to an LSCF thin film membrane ((c) [36]). The relatively poor performance of the 120 µm thin film (hollow blue squares) was due to gas diffusion limitation as the 1 mm thick support had a very low porosity of only 26 %.

The fact that Ba-containing perovskites are prone to carbonate formation in atmospheres containing high amounts of CO_2 [66-69] and in contact with even very low concentrations (cf. Chapter 7, [70]) should be of little consequence if it is used in a three-end module concept. However, X-ray diffraction on annealed powders as well as long term electrical conductivity measurements on dense samples revealed a phase transition to a hexagonal phase at temperatures below 840 °C as well as the formation of another phase with CdI_2 structure (cf. Chapter 8) in the cubic BSCF lattice, thus negatively influencing the electrical performance as well as oxygen permeation.

The reason for this phase transformation seems to be a valence change of the cobalt ions from +2 to +2.7 as measured by Philipp Müller et al. during our cooperation with the Laboratory for Electron Microscopy at the Karlsruhe Institute of Technology [71;72], leading to a decreased ionic radius, in turn destabilizing the cubic lattice. There have been attempts in literature to enhance the stability by complete or partial substitution of the Co by Zn or Al [73;74], reporting enhanced stability, but oxygen permeation was negatively affected. There are, however, indications that relatively small B-site substitutions with Zr [75] or Y show a stabilizing effect and can even – in the case of substitution with 2.5 at% Y – enhance the oxygen flux [76;77].

In this work, measurements were therefore also included concerning BSCF with B-site substitutions by different amounts of Y, Sc and Zr ($(Ba_{0.5}Sr_{0.5})((Co_{0.8}Fe_{0.2})_{1-n/100}X_{n/100})O_{3-\delta}$, further designated BSCFnX, where n denotes the dopant concentration in at% and X stands for the dopant itself) where appropriate.

1.4 Overview

1.4.1 Projects Funding This Work

The measurements and results presented in this work have been performed at the Institut für Werkstoffe der Elektrotechnik (IWE), KIT, in the context of several national and international projects. BSCF has mainly been investigated in the Helmholtz Alliance MEM-BRAIN project[2] and its follow-up, the Helmholtz portfolio project[3], while for the MEM-OXYCOAL[4] project as well as the EU project DEMOYS[5] the focus was more towards LSCF. Thin films have been important for a subproject of the DFG-Center for Functional Nanostructures (CFN)[6].

1.4.2 Chapter Overview

As could be seen in the previous parts of this introduction, there are a lot of material properties important to the understanding and evaluation of a MIEC material. Fig. 1.7 gives a schematic overview of these different aspects.

Electrical conductivity changes with varying oxygen partial pressures and/or temperatures advance the understanding of oxygen stoichiometry changes as well as present indications to possible phase changes in the material during long term measurements. Extraction of transport parameters under different conditions gives information about the material's performance as an oxygen

[2] Helmholtz Alliance MEM-BRAIN: "Gas separation membranes for zero-emission fossil power plants", joint project funded by the German Helmholtz Association of German Research Centers, duration: 2007-2011.

[3] Helmholtz portfolio project "Gasseparationsmembranen für CO_2-freie fossile Kraftwerke", joint project funded by the German Helmholtz Association of German Research Centers, duration: since 2011.

[4] "MEM-OXYCOAL – Sauerstoffpermeable keramische Membranen für kohlebefeuerte Kraftwerke", subproject "Elektrische und elektrochemische Charakterisierungsverfahren", joint project funded by the German Federal Ministry of Education and Research (BMBF) under grant no. 0327803F, duration 2008-2012.

[5] "Dense membranes for efficient oxygen and hydrogen separation (DEMOYS)", joint project funded by the European Commission under the European Community's 7th Framework Programme (Theme 5: Energy), grant agreement no. 241309, duration: 2010-2014.

[6] "Nanostructured functional layers for advanced oxygen separation membranes", subproject F2.1 of the DFG-Center for Functional Nanostructures (CFN) at KIT, funded by the German Research Foundation (Deutsche Forschungsgemeinschaft, DFG), duration: 2009-2013.

ion conductor, while measurements in very low oxygen partial pressures as well as other gases (e.g. CO_2) yield important information in respect to limitations in chemical stability. All these aspects have been the focus of this work, mainly pertaining to BSCF as it has been established as the material of choice in respect to obtainable oxygen permeation rates – which have been measured (as well as mechanical stability under operating conditions) by our partners in the different projects mentioned previously. Results regarding LSCF, PSCF and LSC have been obtained and will be discussed where they present a contribution to the community's knowledge.

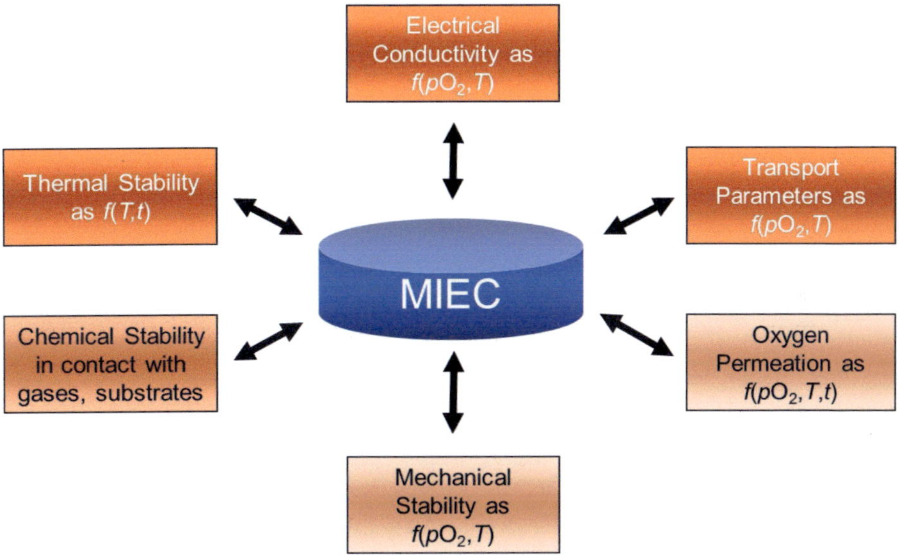

Fig. 1.7: Schematic overview of different aspects important for the understanding of MIEC materials and their performance for possible applications. Saturated colors indicate research done in this work.

In the following, there will be a short overview of what to expect in the different chapters of this work. Starting with Chapter 3, each chapter consists of an introduction to the following measurements, a short experimental section pertaining to the specific chapter, followed by results and discussion as well as the summarized conclusions and the chapter specific literature references.

In this **Chapter 1**, there is a short introduction to the concept of carbon dioxide capture and storage, including an overview over expected drops in efficiency if applied to state-of-the-art coal fired power plants. Different possibilities as to obtain pure CO_2 at the end of a combustion process are shown and the Oxyfuel process, especially with the use of oxygen transport membranes is explained. After a short explanation of two different membrane module concepts, possible perovskite materials are discussed, leading to the focus on BSCF with additional measurements on LSCF, PSCF and LSC – where sensible – in this work.

Chapter 2 contains information pertaining to the measurement setups used for characterization of perovskite powders, bulks and thin films as well as explanations on powder, bulk and thin film (both epitaxial and polycrystalline) preparation.

In **Chapter 3**, the MIEC stability limits at low oxygen partial pressures are analyzed by using the oxygen pump setup. First, the setup is evaluated for the use in coulometric titration measurements (by means of precise background measurements, the use of model metal oxides etc.). BSCF powder is annealed at different partial pressures and analyzed by X-ray diffraction (XRD) measurements in order to find the stability limit of the material at different temperatures. Relative changes in its oxygen stoichiometry are calculated from titration measurements and further titration measurements on LSCF, PSCF, and LSC give indications of the respective stability limits as well as decomposition kinetics of these materials.

Chapter 4 focuses on the conductivity changes in BSCF with changing temperatures. Sintered bulk samples are measured between room temperature and 900 °C in air and pure oxygen. The results are compared to identical measurements on epitaxial and polycrystalline BSCF thin films as well as to measurements on BSCF3X (X = Y, Sc, Zr) bulk samples. Finally, the equilibrium conductivity values are measured at intermediate temperatures, where long dwell-times are necessary to obtain stable values.

In **Chapter 5**, the differences in MIEC conductivity at various oxygen partial pressures indicated in Chapter 4 are examined in greater depth. The oxygen pump setup is used to measure and compare the electrical conductivity of BSCF bulk samples, epitaxial BSCF thin films, sintered LSCF, PSCF and LSC samples and dense BSCF3X (X = Y, Sc, Zr) samples at partial pressures between ambient air and 10^{-5} bar pO_2 and temperatures from 700...900 °C. True equilibrium values for BSCF at 10^{-6} bar and 10^{-8} bar pO_2 at 900 °C could be measured after very long dwell-times of more than 600 hours (due to the vast decrease in oxygen transport parameters at these low partial pressures).

The oxygen pump is used in **Chapter 6** to perform jumps in oxygen partial pressure for electrical conductivity relaxation (ECR) measurements. After a short excerpt concerning possible error sources in such measurements, the mathematical description of the oxygen exchange and diffusion is used to extract surface exchange coefficient values for BSCF, LSCF and PSCF between ambient air pressure and 10^{-5} bar pO_2 at 900 °C, previously not available in literature. A visualization of the total error squares for a large set of k^{δ}-D^{δ}-tupels is used to determine the validity of the fitted values.

Chapter 7 uses ECR in the frequency domain, performed in a kinetic measurement setup on BSCF thin films (both epitaxial and polycrystalline), to extract surface exchange coefficients at intermediate temperatures of 400...700 °C and ambient oxygen partial pressure. The data is compared to each other and – at the higher temperatures – to literature values for bulk samples. Measurements between 400...800 °C on LSCF thin films (sputtered and MOD derived) are used for k^{δ}-value extraction and comparison, at higher temperatures again to literature values. Finally, the technique is used to investigate the deterioration of an epitaxial BSCF thin film sample during repeated measurements in synthetic air containing 0.1 % or 1 % CO_2 at temperatures between 400…600 °C. Post-test analyses by SEM, EDX, and XRD show the destruction of the thin film by carbonate formation.

Starting with SEM/TEM and EDX analyses revealing structural changes and/or secondary phase formation in annealed and post-measurement epitaxial BSCF thin films, **Chapter 8** analyzes the phase stability of BSCF at temperatures from 700…900 °C. Starting with visible degradation of the electrical conductivity of BSCF bulk samples during long term measurements at temperatures below 900 °C – for which measurements a normalization procedure for conductivity changes caused by meteorological air pressure changes had to be applied – the phase transition in BSCF just around the intended operating temperature of 800…850 °C became a concern in respect to future use. Thermal annealing of BSCF powders for increasing amounts of time at 800 °C is presented, yielding XRD derived proof for the slow formation of a hexagonal phase even in pure N_2/O_2 synthetic air. The (reversible) phase transition starts at temperatures below 840 °C with different phase equilibria at different temperatures. Further long term conductivity measurements reveal the phase transition to happen predominantly at grain boundaries. SEM/TEM investigations on annealed and quenched bulk samples – performed by the Laboratory for Electron Microscopy (LEM), KIT, Karlsruhe – are shown and correlate with these electrical measurements, revealing another plate-like phase forming in the system. Annealed samples of BSCF3X (X = Y, Sc, Zr) indicate a stabilizing effect of Y, confirmed in further long term conductivity measurements on BSCFnY (n = 0, 3, and 10) bulk samples, yielding no degradation in conductivity for the samples containing Y.

Finally, the most important results of this thesis are summarized and conclusions drawn – followed by an outlook of ongoing studies pertaining to the BSCF material system for use in OTMs – in **Chapter 9**.

References

[1] Intergovernmental Panel on Climate Change, *The Third Assessment Report of the Intergovernmental Panel on Climate Change*, Cambridge, UK: Cambridge University Press (2001).

[2] Intergovernmental Panel on Climate Change, *Climate Change 2001: the Scientific Basis. Contribution of Working Group I to the Third Assessment Report of the Intergovernmental Panel on Climate Change*, Cambridge, UK: Cambridge University Press (2001).

[3] OECD/IEA, *Energy Balances of Non-OECD Countries, 2001-2002*, Paris: OECD/IEA (2004).

[4] OECD/IEA, *CO_2 emissions from fuel combustion, 1971-2001*, Paris: OECD/IEA (2003).

[5] Intergovernmental Panel on Climate Change, *Climate Change 2001 – Mitigation. The Third Assessment Report of the Intergovernmental Panel on Climate Change*, Cambridge, UK: Cambridge University Press (2001).

[6] Intergovernmental Panel on Climate Change, *IPCC Special Report on Carbon Dioxide Capture and Storage*, Cambridge, United Kingdom and New York, NY, USA: Cambridge University Press (2005).

[7] N. Asprion, *CO_2 Post combustion capture by absorption*, DECHEMA Workshop on CO_2-Capture, -Utilization and -Sequestration (Frankfurt/Main, Germany), 21.01. - 22.01.2008

[8] M. Czyperek, P. Zapp, H. J. M. Bouwmeester, M. Modigell, K. Ebert, I. Voigt, W. A. Meulenberg, L. Singheiser and D. Stöver, "Gas separation membranes for zero-emission fossil power plants: MEM-BRAIN", *Journal of Membrane Science* **359**, pp. 149-159 (2010).

[9] Massachusetts Institute of Technology, *The Future of Coal. Options for a Carbon-Constrained World*, Cambridge: Massachusetts Institute of Technology (2007).

[10] IEA Greenhouse Gas R&D Programme (IEA GHG), "CO_2 capture ready plants, 2007/4", May 2007.

[11] R. Bounaceur, N. Lape, D. Roizard, C. Vallieres and E. Favre, "Membrane processes for post-combustion carbon dioxide capture: A parametric study", *Energy* **31**, pp. 2220-2234 (2006).

[12] E. Favre, "Carbon dioxide recovery from post-combustion processes: can gas permeation membranes compete with absorption?", *Journal of Membrane Science* **294**, pp. 50-59 (2007).

[13] C. Hendriks, *Carbon Dioxide Removal from Coal-fired Power Plants*, Dordrecht/Boston/London: Kluwer Academic Publishers (1994).

[14] A. Hussain and M.-B. Hägg, "A feasibility study of CO_2 capture from flue gas by a facilitated transport membrane", *Journal of Membrane Science* **359**, pp. 140-148 (2010).

[15] T. C. Merkel, H. Lin, X. Wei and R. Baker, "Power plant post-combustion carbon dioxide capture: An opportunity for membranes", *Journal of Membrane Science* **359**, pp. 126-139 (2010).

[16] L. Zhao, E. Riensche, R. Menzer, L. Blum and D. Stolten, "A parametric study of CO_2/N_2 gas separation membrane processes for post-combustion capture", *Journal of Membrane Science* **325**, pp. 284-294 (2008).

[17] L. Zhao, E. Riensche, L. Blum and D. Stolten, "Multistage gas separation membrane processes used in post-combustion capture: energetic and economic analyses", *Journal of Membrane Science* **359**, pp. 160-172 (2010).

[18] A. Corma, F. Ray, J. Rius, M. J. Sabater and S. Valencia, "Supramolecular self-assembled molecules as organic agents directing of synthesis of zeolites", *Nature* **431**, pp. 287-290 (2004).

[19] S. Münzer, J. Caro and P. Behrens, "Preparation and characterization of sodium-free nanocrystalline sodalite", *Microporous and Mesoporous Materials* **110**, pp. 3-10 (2008).

[20] S. Khajavi, F. Kapteijn and J. C. Jansen, "Synthesis of thin defect-free hydroxy sodalite membranes: new candidate for activated water permeation", *Journal of Membrane Science* **299**, pp. 63-72 (2007).

[21] A. van Niekerk, J. Zah, J. C. Breytenbach and H. M. Krieg, "Direct crystallization of hydroxy sodalite membrane without seeding using a conventional oven", *Journal of Membrane Science* **300**, pp. 156-164 (2007).

[22] X. Xu, Y. Bao, C. Song, W. Yang, J. Liu and L. Lin, "Microwave-assisted hydrothermal synthesis of hydroxy-sodalite zeolite membrane", *Microporous and Mesoporous Materials* **75**, pp. 173-181 (2004).

[23] T. Shimura, S. Fujimoto and H. Iwahara, "Protonic conduction in lanthanum strontium aluminate and lanthanum niobate-based oxides at elevated temperatures", *Solid State Ionics* **143**, pp. 117-123 (2001).

[24] R. Haugsrud, "Defects and transport properties in Ln_6WO_{12} (Ln = La, Nd, Gd, Er)", *Solid State Ionics* **178**, pp. 555-560 (2007).

[25] S. Escolástico, V. B. Vert and J. M. Serra, "Preparation and characterization of nanocrystalline mixed proton-electronic conducting materials based on the system Ln_6WO_{12}", *Chemistry of Materials* **21**, pp. 3079-3089 (2009).

[26] J. Davison, "Performance and costs of power plants with capture and storage of CO_2", *Energy* **32**, pp. 1163-1176 (2007).

[27] H. Stadler, F. Beggel, M. Habermehl, B. Persigehl, R. Kneer, M. Modigell and P. Jeschke, "Oxyfuel coal combustion by efficient integration of oxygen transport membranes", *International Journal of Greenhouse Gas Control* **5**, pp. 7-15 (2011).

[28] Y. Teraoka, H. M. Zhang, S. Furukawa and N. Yamazoe, "Oxygen Permeation Through Perovskite-Type Oxides", *Chemistry Letters*, pp. 1743-1746 (1985).

[29] O. Yamamoto, "Low temperature electrolytes and catalysts", in W. Vielstich, A. Lamm, and H. A. Gasteiger (Eds.), *Handbook of Fuel Cells - Vol. 4*, Chichester, England: John Wiley & Sons Ltd., pp. 1002-1014 (2003).

[30] A. Weber and E. Ivers-Tiffée, "Materials and concepts for solid oxide fuel cells (SOFCs) in stationary and mobile applications", *Journal of Power Sources* **127**, pp. 273-283 (2004).

[31] J. Hayd, L. Dieterle, U. Guntow, D. Gerthsen and E. Ivers-Tiffée, "Nanoscaled $La_{0.6}Sr_{0.4}CoO_{3-\delta}$ as intermediate temperature SOFC cathode: Microstructure and electrochemical performance", *Journal of Power Sources* **196**, pp. 7263-7270 (2011).

[32] W. Menesklou, H.-J. Schreiner, K. H. Härdtl and E. Ivers-Tiffée, "High temperature oxygen sensors based on doped $SrTiO_3$", *Sensors and Actuators B-Chemical* **59**, pp. 184-189 (1999).

[33] A. Rothschild, W. Menesklou, H. L. Tuller and E. Ivers-Tiffée, "Electronic structure, defect chemistry, and transport properties of $SrTi_{1-x}Fe_xO_{3-y}$ solid solutions", *Chemistry of Materials* **18**, pp. 3651-3659 (2006).

[34] H. J. M. Bouwmeester and A. J. Burggraaf, "Dense Ceramic Membranes for Oxygen Separation", in P. J. Gellings and H. J. M. Bouwmeester (Eds.), *The CRC Handbook of Solid State Electrochemistry*, Boca Raton, FL: CRC Press, pp. 481-553 (1997).

[35] J. Sunarso, S. Baumann, J. M. Serra, W. A. Meulenberg, Y. S. Lin and J. C. Diniz da Costa, "Mixed ionic-electronic conducting (MIEC) ceramic-based membranes for oxygen separation", *Journal of Membrane Science* **320**, pp. 13-41 (2008).

[36] A. Häffelin, C. Niedrig, S. F. Wagner, S. Baumann, W. A. Meulenberg and E. Ivers-Tiffée, "Three-Dimensional Performance Model for Oxygen Transport Membranes", *Journal of The Electrochemical Society* **161**, pp. F1409-F1415 (2014).

[37] F. Beggel, S. Engels, M. Modigell, and N. Nauels, *OXYCOAL-AC: Integration of High Temperature Membranes for Air Separation into Oxyfuel Power Plants*, 1st Oxyfuel Combustion Conference (Cottbus/Germany), 08.09. - 11.09.2009

[38] H. Kruidhof, H. J. M. Bouwmeester, R. H. E. Vondoorn and A. J. Burggraaf, "Influence of Order-Disorder Transitions on Oxygen Permeability Through Selected Nonstoichiometric Perovskite-Type Oxides", *Solid State Ionics* **63-65**, pp. 816-822 (1993).

[39] S. McIntosh, J. F. Vente, W. G. Haije, D. H. A. Blank and H. J. M. Bouwmeester, "Structure and oxygen stoichiometry of $SrCo_{0.8}Fe_{0.2}O_{3-\delta}$ and $Ba_{0.5}Sr_{0.5}Co_{0.8}Fe_{0.2}O_{3-\delta}$", *Solid State Ionics* **177**, pp. 1737-1742 (2006).

[40] Z. P. Shao, W. S. Yang, Y. Cong, H. Dong, J. H. Tong and G. X. Xiong, "Investigation of the permeation behavior and stability of a $Ba_{0.5}Sr_{0.5}Co_{0.8}Fe_{0.2}O_{3-\delta}$ oxygen membrane", *Journal of Membrane Science* **172**, pp. 177-188 (2000).

[41] S. Baumann, J. M. Serra, M. P. Lobera, S. Escolástico, F. Schulze-Küppers and W. A. Meulenberg, "Ultrahigh oxygen permeation flux through supported $Ba_{0.5}Sr_{0.5}Co_{0.8}Fe_{0.2}O_{3-\delta}$ membranes", *Journal of Membrane Science* **377**, pp. 198-205 (2011).

[42] J. F. Vente, W. G. Haije and Z. S. Rak, "Performance of functional perovskite membranes for oxygen production", *Journal of Membrane Science* **276**, pp. 178-184 (2006).

[43] S. McIntosh, J. F. Vente, W. G. Haije, D. H. A. Blank and H. J. M. Bouwmeester, "Oxygen stoichiometry and chemical expansion of $Ba_{0.5}Sr_{0.5}Co_{0.8}Fe_{0.2}O_{3-\delta}$ measured by in situ neutron diffraction", *Chemistry of Materials* **18**, pp. 2187-2193 (2006).

[44] B. Wei, Z. Lu, X. Q. Huang, J. P. Miao, X. Q. Sha, X. S. Xin and W. H. Su, "Crystal structure, thermal expansion and electrical conductivity of perovskite oxides $Ba_xSr_{1-x}Co_{0.8}Fe_{0.2}O_{3-\delta}$ ($0.3 \leq x \leq 0.7$)", *Journal of the European Ceramic Society* **26**, pp. 2827-2832 (2006).

[45] P. Ried, E. Bucher, W. Preis, W. Sitte and P. Holtappels, "Characterisation of $La_{0.6}Sr_{0.4}Co_{0.2}Fe_{0.8}O_{3-\delta}$ and $Ba_{0.5}Sr_{0.5}Co_{0.8}Fe_{0.2}O_{3-\delta}$ as Cathode Materials for the Application in Intermediate Temperature Fuel Cells", *ECS Transactions* **7**, pp. 1217-1224 (2007).

[46] W. Zhou, R. Ran, Z. P. Shao, W. Zhuang, J. Jia, H. X. Gu, W. Q. Jin and N. P. Xu, "Barium- and strontium-enriched $(Ba_{0.5}Sr_{0.5})_{(1+x)}Co_{0.8}Fe_{0.2}O_{3-\delta}$ oxides as high-performance cathodes for intermediate-temperature solid-oxide fuel cells", *Acta Materialia* **56**, pp. 2687-2698 (2008).

[47] E. Girdauskaite, H. Ullmann, V. V. Vashook, U. Guth, G. B. Caraman, E. Bucher and W. Sitte, "Oxygen transport properties of $Ba_{0.5}Sr_{0.5}Co_{0.8}Fe_{0.2}O_{3-x}$ and $Ca_{0.5}Sr_{0.5}Mn_{0.8}Fe_{0.2}O_{3-x}$ obtained from permeation and conductivity relaxation experiments", *Solid State Ionics* **179**, pp. 385-392 (2008).

[48] E. Bucher, A. Egger, P. Ried, W. Sitte and P. Holtappels, "Oxygen nonstoichiometry and exchange kinetics of $Ba_{0.5}Sr_{0.5}Co_{0.8}Fe_{0.2}O_{3-\delta}$", *Solid State Ionics* **179**, pp. 1032-1035 (2008).

[49] L. Wang, R. Merkle, J. Maier, T. Acarturk and U. Starke, "Oxygen tracer diffusion in dense $Ba_{0.5}Sr_{0.5}Co_{0.8}Fe_{0.2}O_{3-\delta}$ films", *Applied Physics Letters* **94**, pp. 071908-1-071908-3 (2009).

[50] M. Burriel, C. Niedrig, W. Menesklou, S. F. Wagner, J. Santiso and E. Ivers-Tiffée, "BSCF epitaxial thin films: Electrical transport and oxygen surface exchange", *Solid State Ionics* **181**, pp. 602-608 (2010).

[51] C. Niedrig, W. Menesklou, S. F. Wagner and E. Ivers-Tiffée, "High-Temperature pO_2 Stability of Metal Oxides Determined by Amperometric Oxygen Titration", *Journal of The Electrochemical Society* **160**, pp. F135-F140 (2013).

[52] J. E. ten Elshof, M. H. R. Lankhorst and H. J. M. Bouwmeester, "Chemical diffusion and oxygen exchange of $La_{0.6}Sr_{0.4}Co_{0.6}Fe_{0.4}O_{3-\delta}$", *Solid State Ionics* **99**, pp. 15-22 (1997).

[53] J. A. Lane, S. J. Benson, D. Waller and J. A. Kilner, "Oxygen transport in $La_{0.6}Sr_{0.4}Co_{0.2}Fe_{0.8}O_{3-\delta}$", *Solid State Ionics* **121**, pp. 201-208 (1999).

[54] G. C. Kostogloudis and C. Ftikos, "Properties of A-site-deficient $La_{0.6}Sr_{0.4}Co_{0.2}Fe_{0.8}O_{3-\delta}$-based perovskite oxides", *Solid State Ionics* **126**, pp. 143-151 (1999).

[55] S. Wang, P. A. W. van der Heide, C. Chavez, A. J. Jacobson and S. B. Adler, "An electrical conductivity relaxation study of $La_{0.6}Sr_{0.4}Fe_{0.8}Co_{0.2}O_{3-\delta}$", *Solid State Ionics* **156**, pp. 201-208 (2003).

[56] H. J. M. Bouwmeester, M. W. den Otter and B. A. Boukamp, "Oxygen transport in $La_{0.6}Sr_{0.4}Co_{1-y}Fe_{y}O_{3-\delta}$", *Journal of Solid State Electrochemistry* **8**, pp. 599-605 (2004).

[57] A. Leonide, B. Rüger, A. Weber, W. A. Meulenberg and E. Ivers-Tiffée, "Impedance Study of Alternative $(La,Sr)FeO_{3-\delta}$ and $(La,Sr)(Co,Fe)O_{3-\delta}$ MIEC Cathode Compositions", *Journal of The Electrochemical Society* **157**, pp. B234-B239 (2010).

[58] V. A. C. Haanappel, A. Mai, S. Uhlenbruck and F. Tietz, "Characterization of Anode-Supported Solid Oxide Fuel Cells With PSCF Cathode", *Journal of Fuel Cell Science and Technology* **6**, pp. 1-6 (2009).

[59] C. S. Barclay, L. Zhao, S. R. Bishop and K. L. Duncan, "Defect Equilibria of $(Pr_{0.6}Sr_{0.4})(Co_{0.2}Fe_{0.8})O_{3-d}$", *ECS Transactions* **57**, pp. 2105-2113 (2013).

[60] J. Mizusaki, Y. Mima, S. Yamauchi, K. Fueki and H. Tagawa, "Nonstoichiometry of the perovskite-type oxides $La_{1-x}Sr_{x}CoO_{3-\delta}$", *Journal of Solid State Chemistry* **80**, pp. 102-111 (1989).

[61] M. H. R. Lankhorst, H. J. M. Bouwmeester and H. Verweij, "High-Temperature Coulometric Titration of $La_{1-x}Sr_{x}CoO_{3-\delta}$: Evidence for the Effect of Electronic Band Structure on Nonstoichiometry Behavior", *Journal of Solid State Chemistry* **133**, pp. 555-567 (1997).

[62] S. B. Adler, "Mechanism and kinetics of oxygen reduction on porous $La_{1-x}Sr_{x}CoO_{3-\delta}$ electrodes", *Solid State Ionics* **111**, pp. 125-134 (1998).

[63] M. Søgaard, P. V. Hendriksen, M. Mogensen, F. W. Poulsen and E. Skou, "Oxygen nonstoichiometry and transport properties of strontium substituted lanthanum cobaltite", *Solid State Ionics* **177**, pp. 3285-3296 (2006).

[64] J. Hayd, H. Yokokawa and E. Ivers-Tiffée, "Hetero-Interfaces at Nanoscaled $(La,Sr)CoO_{3-\delta}$ Thin-Film Cathodes Enhancing Oxygen Surface-Exchange Properties", *Journal of The Electrochemical Society* **160**, pp. F351-F359 (2013).

[65] M. Betz, F. Schulze-Küppers, S. Baumann, W. A. Meulenberg and D. Stöver, "Supported oxygen transport membranes for oxyfuel power plants", *Advances in Science and Technology* **72**, pp. 93-98 (2010).

[66] M. Arnold, H. Wang and A. Feldhoff, "Influence of CO_2 on the oxygen permeation performance and the microstructure of perovskite-type $(Ba_{0.5}Sr_{0.5})(Co_{0.8}Fe_{0.2})O_{3-\delta}$ membranes", *Journal of Membrane Science* **293**, pp. 44-52 (2007).

[67] S. A. Möbius, *Charakterisierung perowskitischer Hochtemperaturmembranen zur Sauerstoffbereitstellung für fossil gefeuerte Kraftwerksprozesse*, Dissertation, Fakultät für Maschinenwesen, RWTH Aachen (2010).

[68] A. Yan, M. Cheng, Y. L. Dong, W. S. Yang, V. Maragou, S. Q. Song and P. Tsiakaras, "Investigation of a $Ba_{0.5}Sr_{0.5}Co_{0.8}Fe_{0.2}O_{3-\delta}$ based cathode IT-SOFC I. The effect of CO_2 on the cell performance", *Applied Catalysis B-Environmental* **66**, pp. 64-71 (2006).

[69] A. Waindich, A. Möbius and M. Müller, "Corrosion of $Ba_{1-x}Sr_xCo_{1-y}Fe_yO_{3-\delta}$ and $La_{0.3}Ba_{0.7}Co_{0.2}Fe_{0.8}O_{3-\delta}$ materials for oxygen separating membranes under Oxycoal conditions", *Journal of Membrane Science* **337**, pp. 182-187 (2009).

[70] E. Bucher, A. Egger, G. B. Caraman and W. Sitte, "Stability of the SOFC cathode material $(Ba,Sr)(Co,Fe)O_{3-\delta}$ in CO_2-containing atmospheres", *Journal of The Electrochemical Society* **155**, pp. B1218-B1224 (2008).

[71] P. Müller, H. Störmer, M. Meffert, L. Dieterle, C. Niedrig, S. F. Wagner, E. Ivers-Tiffée and D. Gerthsen, "Secondary phase formation in $Ba_{0.5}Sr_{0.5}Co_{0.8}Fe_{0.2}O_{3-\delta}$ studied by electron microscopy", *Chemistry of Materials* **25**, pp. 564-573 (2013).

[72] P. Müller, M. Meffert, H. Störmer and D. Gerthsen, "Fast Mapping of the Cobalt-Valence State in $Ba_{0.5}Sr_{0.5}Co_{0.8}Fe_{0.2}O_{3-d}$ by Electron Energy Loss Spectroscopy", *Microscopy and Microanalysis* **19**, pp. 1595-1605 (2013).

[73] A. Feldhoff, J. Martynczuk and H. Wang, "Advanced $Ba_{0.5}Sr_{0.5}Zn_{0.2}Fe_{0.8}O_{3-\delta}$ perovskite-type ceramics as oxygen selective membranes: Evaluation of the synthetic process", *Progress in Solid State Chemistry* **35**, pp. 339-353 (2007).

[74] J. Martynczuk, K. Efimov, L. Robben and A. Feldhoff, "Performance of zinc-doped perovskite-type membranes at intermediate temperatures for long-term oxygen permeation and under a carbon dioxide atmosphere", *Journal of Membrane Science* **344**, pp. 62-70 (2009).

[75] S. Yakovlev, C.-Y. Yoo, S. Fang and H. J. M. Bouwmeester, "Phase transformation and oxygen equilibration kinetics of pure and Zr-doped $Ba_{0.5}Sr_{0.5}Co_{0.8}Fe_{0.2}O_{3-\delta}$ perovskite oxide probed by electrical conductivity relaxation", *Applied Physics Letters* **96**, pp. 1-3 (2010).

[76] P. Haworth, S. Smart, J. Glasscock and J. C. Diniz da Costa, "Yttrium doped BSCF membranes for oxygen separation", *Separation and Purification Technology* **81**, pp. 88-93 (2011).

[77] P. Haworth, S. Smart, J. Glasscock and J. C. Diniz da Costa, "High performance yttrium-doped BSCF hollow fibre membranes", *Separation and Purification Technology* **94**, pp. 16-22 (2012).

Chapter 2

Experimental

2.1 Structural Characterization

2.1.1 X-Ray Diffractometry

Unless indicated otherwise, all X-ray diffractometrical (XRD) analyses on powders and bulks were carried out using a Siemens D-5000 spectrometer (Bruker-AXS, Karlsruhe, Germany) with CuK_{α} radiation and a scintillation detector. The measurements were performed in a 2Θ-range from 20° to 80° with a current and voltage of 40 mA and 40 kV respectively, step size was set to 0.01° with a scan speed of 10 seconds per step in the locked coupled mode and a variable slit (V20) for both divergence and antiscatter slit as well as a Ni filter with a thickness of 0.2 mm in front of the detector. For measurements on thin films the XRD setup was operated in the grazing incidence mode with a tube angle of 1° and a monochromator (Bruker-AXS) on the detector side.

All spectra have been corrected for $CuK_{\alpha 2}$-radiation and the background has been subtracted, using the program EVA (Bruker AXS) unless indicated otherwise.

2.1.2 Particle Size Analyses

The particle size analysis of powders was performed using a CILAS 1064L particle sizer. By laser light diffraction pattern analysis, particle size distribution can be measured in the range from 0.04 to 500 μm.

2.1.3 SEM/EDX Analyses

Scanning electron microscopy images have been recorded using a LEO 1530 (Zeiss, Oberkochen, Germany) scanning electron microscope (SEM). Acceleration voltage (EHT), working distance (WD) and the selected detector (Signal A) are indicated in the footer of every image together with displayed magnification (Mag).

Elemental chemical analysis of the BSCF powder was performed by Energy-Dispersive X-ray Analysis (EDX) performed with an INCAx-SIGHT detector from Oxford Instruments (measurement in Fig. 2.7) and an XFlash Detector 5030 from Bruker AXS (rest).

SEM/EDX and TEM analyses shown in the second part of Chapter 8 have been performed by P. Müller at the Laboratory for Electron Microscopy (LEM) of the Karlsruhe Institute of Technology (KIT), Germany. Specifics are given in that chapter.

2.2 Thermal Treatment

Calcination, sintering, firing of the electrical contacts, and thermal annealing of the MIEC powders and bulks in ambient air was done using chamber furnaces, while treatment in synthetic air was performed in a tube furnace with a mixture of 20 % oxygen in N_2. Heating and cooling rates have been fixed to 20 K/min except where indicated otherwise and for the bulk samples examined by the LEM in Chapter 8, which were directly quenched in water after the intended annealing time had been reached. For annealing experiments in low-pO_2 atmospheres cf. section 2.3.2.

2.3 Electrical, Kinetic and Coulometric Titration Measurements

2.3.1 Long Term Conductivity Measurements

All long term conductivity measurements conducted on sintered bulk samples have been performed in ambient air in a chamber furnace. This setup allows for five four-wire contacted samples to be measured at the same time by applying a serial current using a Keithley 224 programmable current source while sequentially reading the voltage over each sample's contacts using a Keithley 2000 multimeter. The possibility of sample degradation over time due to kinetic ionic demixing (cation migration) as a result of long term DC loading could be ruled out by a careful comparison of two different long term measurements on the same set of (new) samples, one with a continuous current, the other with a setting that only applied a measurement current for one minute every hour to get a set of data points. Both experiments yielded the same rate of degradation, ruling out kinetic demixing as a possible error source. Given the fairly high conductivity of the investigated MIEC materials and the resulting small electric fields in the sample, this was to be expected.

2.3.2 Oxygen Pump Setup

The "oxygen pump" is used extensively in this work. With it, electrical conductivity measurements at very low oxygen partial pressures can be performed as well as low-pO_2 annealing experiments and electrical conductivity relaxation (ECR) measurements on dense MIEC bulk samples. Even stability experiments on powders and measurements of stoichiometry changes with oxygen partial pressure become available by means of coulometric titration. Chapters 3 and 6 contain an in-depth validation of the setup's capabilities concerning coulometric titration and ECR, respectively. The setup was introduced by Beetz [1] for pO_2 control and has been modified during this work for titration and stability investigations (cf. chapter 3, [2;3]). It consists of a sealed and gas-tight zirconia (YSZ) tube, containing the sample space (Fig. 2.1). After heating the zirconia tube to temperatures above approx. 500 °C, the intrinsic oxygen ion conductivity of YSZ enables oxygen transport into or out of the sample chamber, driven by an applied voltage U_{pump} between the large-area porous platinum pump-electrodes on the outside and inside of the YSZ-tube. At temperatures above approx. 700 °C, oxygen transport becomes sufficiently fast for electrical conductivity and titration measurements as described in later chapters.

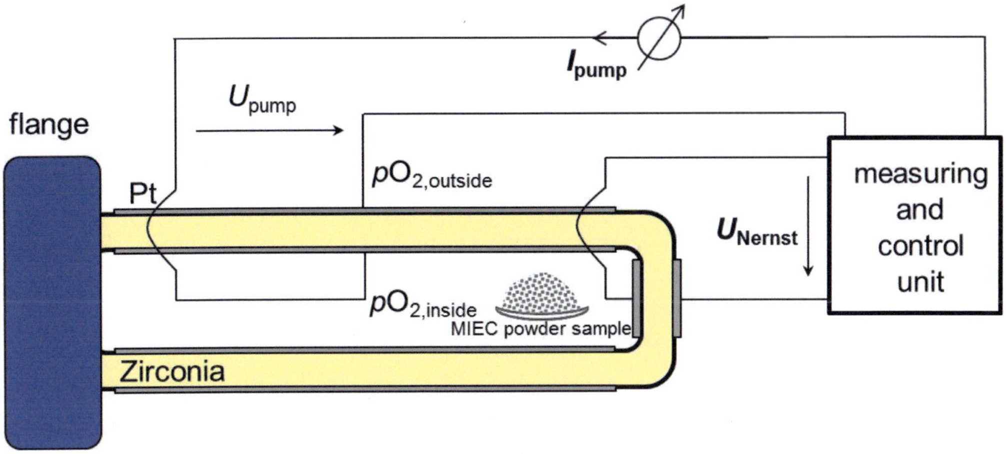

Fig. 2.1: Schematic of the "oxygen pump" measurement setup with a powder sample.

While flushing the outside of the tube continuously with pure oxygen or compressed air – and therefore providing a constant and well known $pO_{2,outside}$ at the outer Pt-electrodes – the application of the voltage U_{pump} leads to the transport of oxygen ions through the YSZ lattice into or out of the sample chamber. Linearly increasing the voltage (at a well-defined rate, typically 1 mV/min) then results in an exponential decrease of the pO_2 in the sample chamber. By constantly monitoring the Nernst voltage U_{Nernst} between the reference electrodes very close to the sample (cf. Fig. 2.1), the oxygen partial pressure at the sample's position can be precisely adjusted in the range from 10^{-18}...1 bar using eq. (2.1) for the Nernst voltage

$$U_{Nernst} = \frac{RT}{4F} \ln \frac{pO_{2,outside}}{pO_{2,inside}} \tag{2.1}$$

between sample chamber (volume ~38 cm³) and outer gas compartment, where a constant gas flow maintains $pO_{2,outside}$. F is Faraday's constant, R the universal gas constant, T the absolute temperature, and $pO_{2,outside}$ and $pO_{2,inside}$ are the oxygen partial pressures outside and inside the sample chamber, respectively. The electronic setup to control pO_2 and measure the pump current is custom-made at our institute (IWE).

In order to prevent any influence of atmospheric gases, e.g. N_2, CO_2 and other trace gases in ambient air, at the beginning of each measurement the sample chamber is flushed with pure oxygen (99.95 % purity) and then sealed using water-cooled KF flanges at the low-temperature end of the zirconia tube.

This "oxygen pump" is used for electrical conductivity measurements on MIEC bulk samples at constant oxygen partial pressures, ECR measurements on bulk samples by performing step changes in pO_2, annealing of powders at well-defined single pO_2 values by simply exposing the powder sample (approximately 0.5 g per measurement) to the preset atmosphere within the tube and coulometric titration of oxide powder samples (located at the sample position indicated in Fig. 2.1 inside a Pt crucible) at a set pO_2 profile over time. Electrical measurements on bulk samples were carried out using an Agilent Micro Ohm Meter (34420 A) and four-wire contacts.

2.3.3 Kinetic Setup

ECR measurements can also be performed in the frequency domain, where the fast kinetic behavior of samples is investigated in a pressure modulation setup first presented by Tragut et al. [4], as shown in a schematic overview in Fig. 2.2. The total pressure p of compressed synthetic air – ($pO_2 = xO_2 \cdot p$, $xO_2 = 0.21$) – or other chosen synthetic gas compositions provided in commercial gas cylinders (Air Liquide) – in the sample space is modulated between ambient air pressure and 1.1…1.3 bar by means of alternately activated magnetic valves, resulting in periodical pO_2 changes with a frequency f up to ~ 30 Hz. By staying close to the electrochemical equilibrium of the system due to the small pressure variations, the behavior can be described by a first-order model. While monitoring the pressure modulation with a reference sensor, the corresponding changes in electrical conductance $G(t)$ are measured as a function of $pO_2(t)$, thereby determining the amplitude and phase shift of both excitation signal ($pO_2(t)$) and its response ($G(t)$); after a Fourier transformation, the frequency response can be analyzed. For surface-controlled samples, easily recognizable by the slope of – 1 in the double-logarithmic Bode plot (cf. Fig. 2.2 bottom right), the surface exchange parameter k^δ can then be determined according to a model described in more detail in [4;5]:

$$G(i\omega) = G_1 \sinh\left(\frac{d}{2}\Omega\right) \cdot \left[\Omega \cdot \cosh\left(\frac{d}{2}\Omega\right) + \frac{i\omega}{k^\delta} \cdot \sinh\left(\frac{d}{2}\Omega\right)\right]^{-1} \tag{2.1}$$

Where d is the sample thickness, i the imaginary unit, $\omega = 2\pi f$, $\Omega = \sqrt{i\omega / D^\delta}$ and G_1 a constant without further significance.

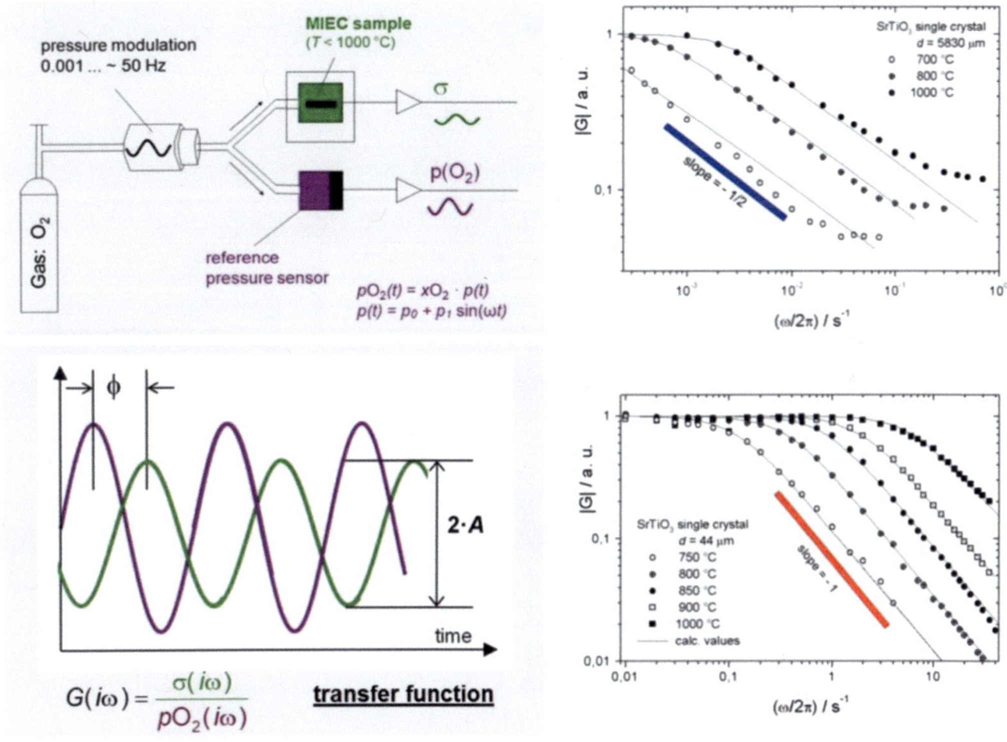

Fig. 2.2: Schematic of conductivity relaxation measurements in the frequency domain (top left), the derived transfer function (bottom left) and measurements on $SrTiO_3$ single crystals with typical diffusion controlled (top right) and surface exchange controlled (bottom right) behavior [5].

2.4 Sample Preparation

2.4.1 MIEC Powders

In this work, MIEC powders have been used for direct measurements (chemical stability limit determination in the oxygen pump setup, stoichiometry change examinations by coulometric titration, thermal stability measurements by annealing) as well as for the preparation of sintered bulk samples for structural examination and electrical measurements, and the production of targets for the preparation of epitaxial thin films by pulsed laser deposition (PLD).

For the measurements on powders shown in this work, commercially available (Treibacher, Austria) $Ba_{0.5}Sr_{0.5}Co_{0.8}Fe_{0.2}O_{3-\delta}$ (BSCF) and $((Ba_{0.5}Sr_{0.5})((Co_{0.8}Fe_{0.2})_{1-n/100}X_{n/100})O_{3-\delta}$, (BSCFnX, n = 3, X = Y, Sc, Zr) have been provided in the context of the MEM-BRAIN project to all research partners – in order to guarantee comparability of the results – by the Fraunhofer Institute for Ceramic Technologies and Systems (IKTS), Hermsdorf/Germany. $La_{0.58}Sr_{0.4}Co_{0.2}Fe_{0.8}O_{3-\delta}$ (LSCF) and $Pr_{0.58}Sr_{0.4}Co_{0.2}Fe_{0.8}O_{3-\delta}$ (PSCF) have been prepared by spray pyrolysis and kindly provided by Forschungszentrum Jülich, Institute of Energy and Climate Research (IEK-1), Jülich, Germany.

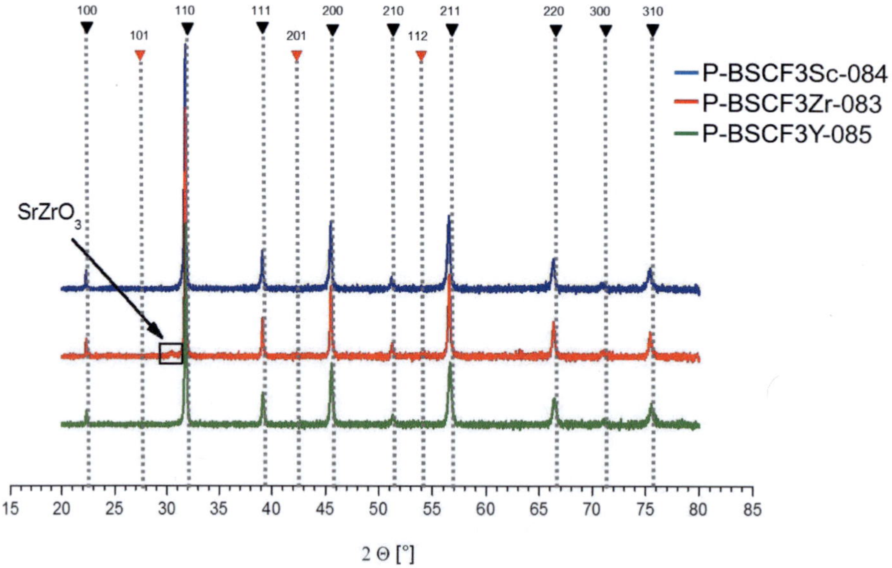

Fig. 2.3: XRD patterns of doped BSCF powders provided by IKTS. Indicated are the positions of cubic (black triangles) and hexagonal (red triangles) perovskite reflections. A secondary $SrZrO_3$ phase was detected in P-BSCF3Zr-083.

$La_{0.6}Sr_{0.4}CoO_{3-\delta}$ (LSC), the BSCF powder used for the preparation of PLD targets, as well as BSCF10Y have been prepared at IWE by the mixed-oxide route. X-ray diffractometry (XRD) analyses confirmed all raw powders to be single-phase perovskite compositions after preparation, except for BSCF3Zr, where $SrZrO_3$ was detected (Fig. 2.3). The D_{50} value of the powders was determined by particle-sizer analysis to be around 2…3 µm for BSCF, LSCF, PSCF and LSC and around 6…7 µm for BSCF3X. The XRD analysis of the BSCF powder (D_{50} = 2.40 µm) revealed a lattice constant of 3.986 Å of the single phase cubic perovskite structure, which corresponds well to the values determined by other groups, e.g. [6] and [7]. All nine reflections in the 2Θ-range from 20° to 80° (cf. graph (a) in Fig. 8.10 in Chapter 8) have been fitted using WinINDEX (Bruker AXS), yielding only one possible result.

Given the fact that the BSCF system is the main focus of this work, as well as the necessity of powder preparation to achieve different levels of doping with Y, Sc or Zr, powder preparation has been performed in order to gain insights into calcination times, temperatures and milling procedures. Fig. 2.4 shows a schematic detailing the mixed-oxide route used for the BSCF powder preparation. Stoichiometric amounts of $BaCO_3$, $SrCO_3$, Co_3O_4 and Fe_2O_3 powders from MERCK (Darmstadt, Germany) are mixed together for 24 hours on a rolling bench (100 rpm), then calcined for 10 hours at different temperatures, followed by milling with zirconia balls (again on the rolling bench).

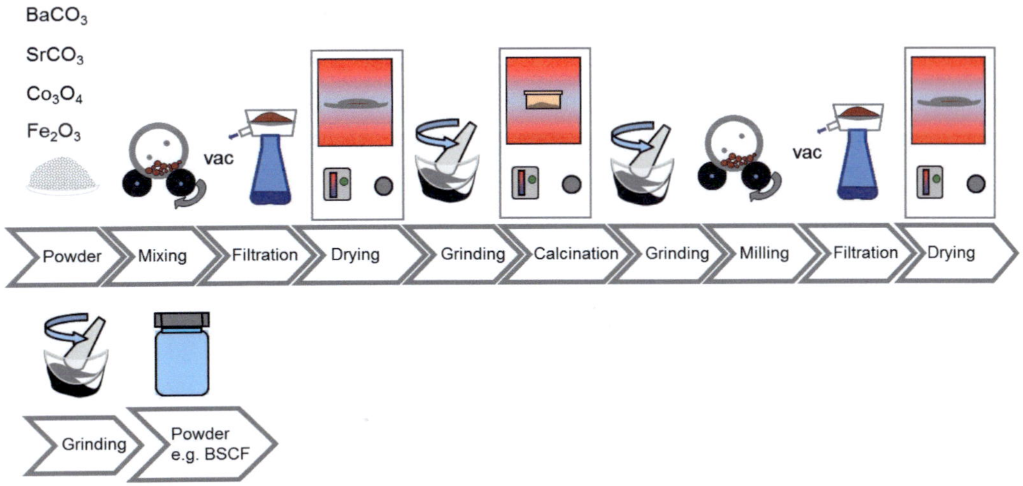

Fig. 2.4: Mixed-oxide route for powder preparation from oxides and carbonates (schematic based on [8]).

For the calcined powders to consist of the single phase cubic BSCF perovskite, the annealing temperature had to be at least 950 °C, as can be seen in the XRD measurements shown in Fig. 2.5.

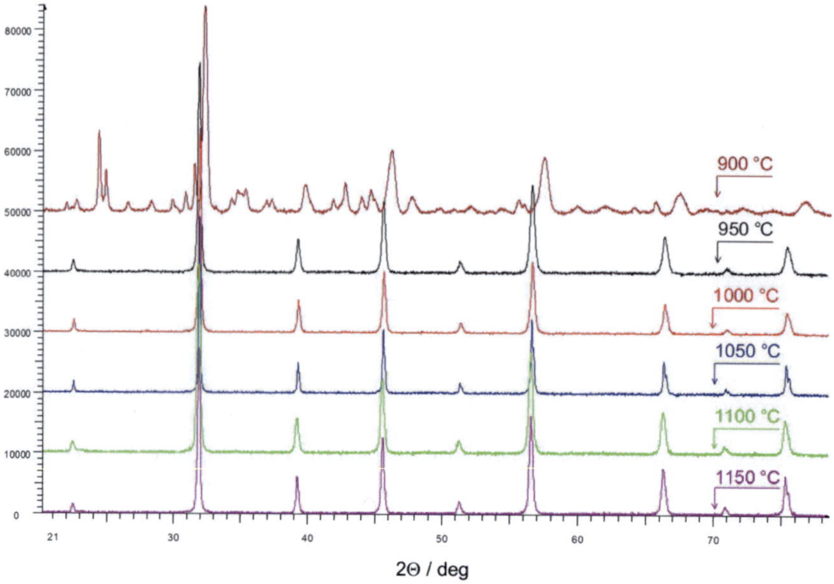

Fig. 2.5: XRD-patterns of BSCF powders calcined at different temperatures. Single phase cubic perovskite is obtained at temperatures above 900 °C.

The powder calcined for 10 hours at 900 °C can still become single phase cubic BSCF after an additional calcination for 20 more hours at the same temperature. This possibility becomes more important as milling experiments showed the freshly produced BSCF to be very hard, therefore representing a challenge in respect to achieving suitably small particle sizes by ball milling. A powder calcined for 10 hours at 1050 °C needed 72 hours of milling time to be ground down to a D_{50} value of 3.26 μm with a still slightly bimodal particle size distribution. However, these long milling procedures are problematic due to the introduction of zirconia into the powders from the milling balls.

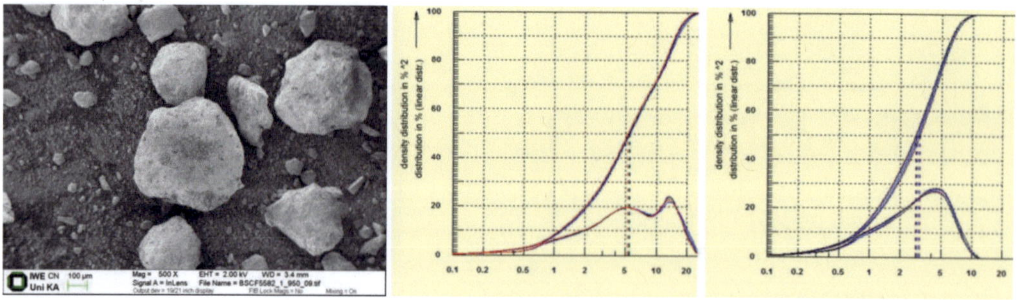

Fig. 2.6: SEM-picture of powder IWE-BSCF-03-0100 after calcination at 950 °C (left, green length indicator is 100 μm) and grain size distribution after 5 hours of milling (middle, D_{50} = 5.26 μm) and 34 hours of milling (right, D_{50} = 3.09 μm).

However, lower calcination temperatures result in powders that are somewhat easier to grind down, as can be seen in Fig. 2.6. After 34 hours of milling, a nearly monomodal particle size distribution with a D_{50} value of 3.09 µm can be achieved. A nearly similar distribution with a D_{50} value of 3.22 µm can be obtained by milling for 9 hours in a planetary mill, but resulting in higher contamination with zirconia. Calcination at 900 °C and subsequent milling for 44 hours on the rolling bench resulted in a monomodal particle size distribution and a D_{50} value of 1.7 µm. This mixed-oxide route can easily be modified by adding oxides of Y, Sc or Zr in the desired stoichiometric amount during the first step in order to obtain doped BSCFnX powders.

Fig. 2.7 shows analyses performed on the provided BSCF powder, as it is used throughout this work for the powder analyses and – sintered to bulk samples – for the electrical and stability measurements.

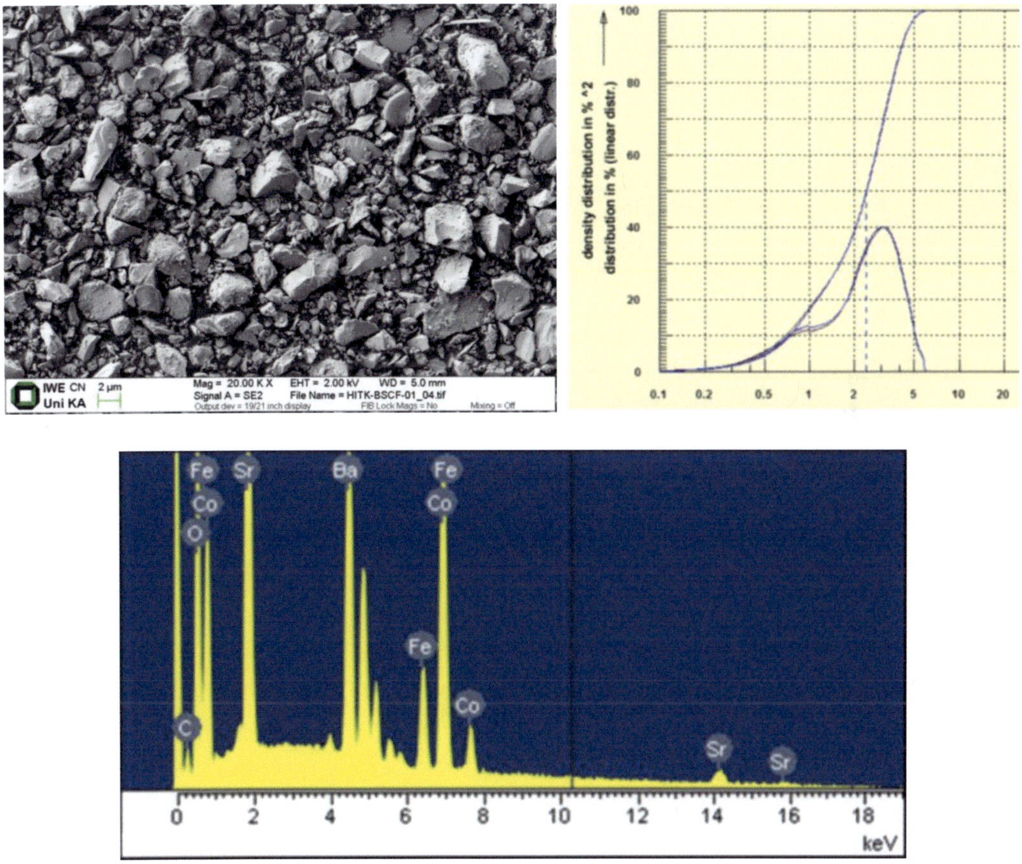

Fig. 2.7: SEM picture and grain size distribution of BSCF powder provided by IKTS (top, $D_{50} = 2.40$ µm) and EDX spectrum of the powder, yielding a stoichiometry of $Ba_{0.50}Sr_{0.51...0.60}Co_{0.72...0.73}Fe_{0.19}O_{3-\delta}$ (by normalizing to 50 % Ba on the A site).

The rugged form of the particles suggests aggressive milling has been applied during the production process, which, however, did not pose any problems during bulk preparation. EDX analysis shows a good stoichiometric fit to the intended values, except for a slightly lower Co-content.

2.4.2 Sintered MIEC Bulk Samples

As there are a variety of different sintered samples used for measurements in this work, a list of samples provided in the Appendix shows the relevant production conditions in an easily accessible overview. In general, however, the different powders (cf. Section 2.4.1) were used to produce dense ceramic bulk samples. Most of the BSCF, LSCF and PSCF bulk samples shown in this work (cf. list of samples in the Appendix) were prepared and supplied in the context of the MEM-BRAIN and MEM-OXYCOAL projects (cf. Chapter 1) by Forschungszentrum Jülich GmbH, IEK-1, Jülich/Germany. Preparation consisted of uniaxial pressing at approx. 10 kN/cm² and subsequent sintering. For these standardized samples the sintering conditions were slightly different between the different materials: in the case of BSCF, sintering occurred for 12 hours at 1000 °C, in the case of LSCF for 5 h at 1200 °C, while PSCF was sintered for 5 h at 1250 °C. This resulted in very dense pellets (SEM cross-sectional analyses showed densities of > 95 %), that were subsequently cut by ultrasonic lapping and mechanically polished (P1200 emery paper). All LSC bulks were prepared at IWE by isostatic pressing at approx. 20 kN/cm², subsequent sintering for 12 h at 1350 °C and mechanical surface polishing. Heating and cooling rates were kept at 5 K/min if not indicated otherwise.

In order to be able to produce sintered samples from BSCF powders prepared at IWE, the BSCF3X (X = Y, Sc or Zr) powders supplied by IKTS Hermsdorf (cf. section 2.4.1.) and self-prepared powders with different contents of Y, it is however prudent to know about the sintering behavior of the different materials. To this end, several series of sintering experiments have been conducted. All bulk samples prepared at IWE have been produced by isostatic pressing with 27 kN/cm² and subsequent sintering.

Fig. 2.8 illustrates the influence of powder grain size on the resulting bulk density, by comparing two SEM micrographs of BSCF bulks sintered for 10 hours at 1000 °C from powders with D_{50} values of 6.04 μm (left) and 3.26 μm (right). It can clearly be seen that much higher densities can be achieved by using powders with smaller grain sizes. This becomes increasingly evident during the sintering experiments on the provided BSCF3X powders shown later in this section, as their respective D_{50} values range between 6..7 μm.

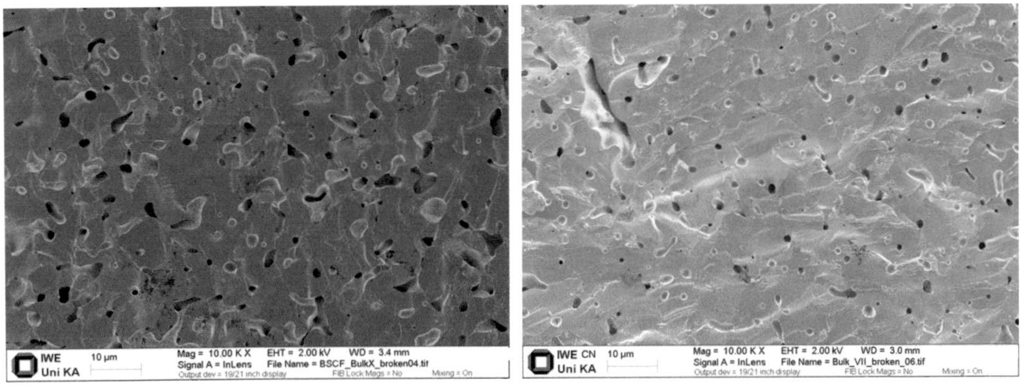

Fig. 2.8: SEM pictures of BSCF samples sintered for 10 hours at 1000 °C from powders with different grain sizes (6.04 µm, left and 3.26 µm, right).

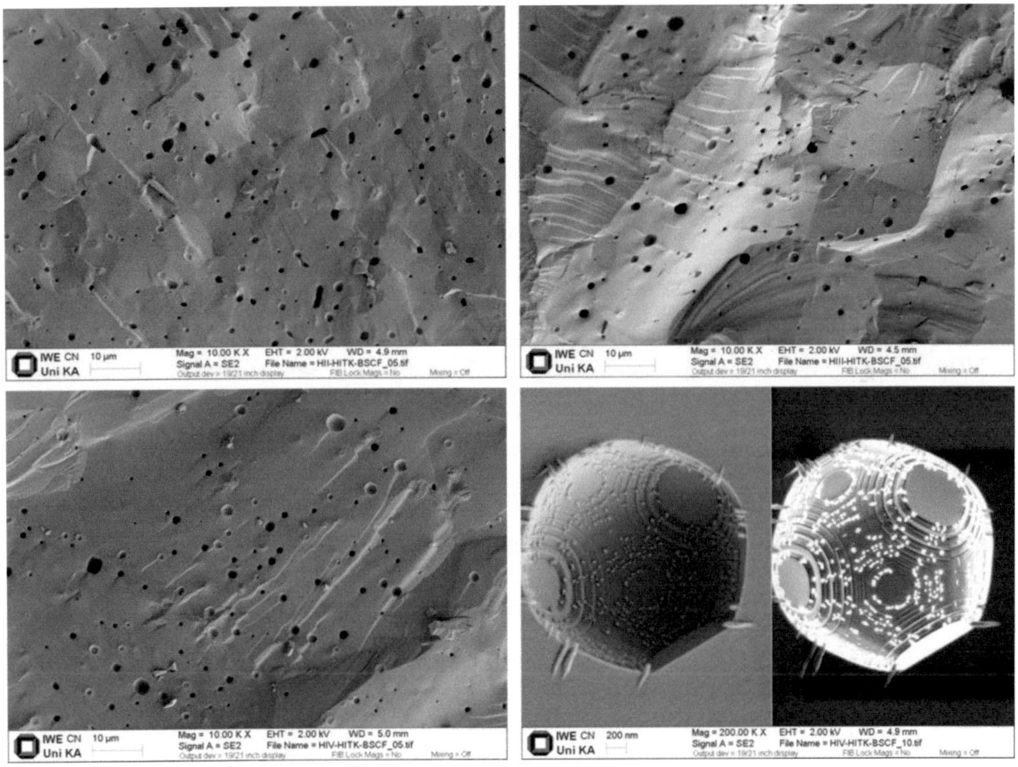

Fig. 2.9: SEM pictures of BSCF samples sintered from IKTS powder at different temperatures for different times (1000 °C / 20 h: top left, 1100 °C / 10 h: top right, 1100 °C / 20 h: bottom left) and a pore with nanoscaled condensed secondary phases (bottom right).

For the sintering behavior of the IKTS-supplied BSCF powder, Fig. 2.9 shows a comparison of cross section SEM micrographs taken from bulk samples produced under different sintering conditions (*T*, *t*). The achievable density is quite comparable between bulks sintered at 1000 °C for 20 hours (top left) and those sintered at 1100 °C for 10 hours (top right). However, sintering at 1100 °C for 20 hours (bottom left) could lead to an ever higher density, but the examination also suggests the temperature to be dangerously close to the melting temperature of BSCF (approx. 1180 °C, [9]), manifesting this fact in strange spherical pores decorated with nano-scale secondary phase precipitations (bottom right), probably early indications of liquid phase sintering. Therefore, sintering temperatures for BSCF bulks have been generally kept at 1000 °C both at IEK-1 and IWE.

An exception was made in order to achieve bulk samples with different grain sizes for the long term measurements shown in Fig. 8.17 in Chapter 8, which have been produced by IEK-1 at the 'standard' conditions of 1000 °C/12 h for resulting grain sizes of approx. 25 μm, as well as higher sintering temperatures and longer times of 1120 °C/50 h and 1150 °C/50 h for mean grain sizes around 50 μm and 90 μm, respectively, as determined by analyses using the software SPIP (Image Metrology A/S, Denmark).

In the case of producing sintered bulk samples from the provided BSCF3X powders, optimal sintering conditions had to be determined due to the fact of increased powder grain sizes (6...7 μm) as well as possibly different sintering behavior of the doped BSCF system. Fig. 2.10 shows an overview of cross-section SEM-micrographs taken from BSCF3X (X = Y, Sc or Zr) bulks pressed isostatically with 27 kN/cm² and sintered for 12 hours at 1050 °C, 1100 °C and 1150 °C. The lowest porosities (albeit much higher than those in standard BSCF samples, due to the larger powder grain sizes) of 11 %, 15 % and 14 % could be achieved at 1100 °C for the Sc- and Zr-containing samples and 1150 °C for the Y-containing ones, respectively. These sintering conditions – marked with the red borders in Fig. 2.10 – were therefore used in the production of samples for electrical measurements in this work (cf. list of samples in the Appendix for data on specific samples).

EDX analyses on BSCF3X samples, however, revealed secondary phases mainly consisting of the used dopant in all cases, as shown in Fig. 2.11 on the example of a BSCF3Y bulk cross-section. This effectively lowers the dopant concentration in the main BSCF matrix to a value below the nominal 3 at% and has to be kept in mind in regard to the findings presented on the samples made from these powders. These analyses have been confirmed by measurements conducted by P. Müller and M. Meffert at the Laboratory for Electron Microscopy (LEM) at KIT. In-depth structural analyses of these samples, along with those performed on BSCF samples, are shown in detail in the pertaining sections of Chapter 8.

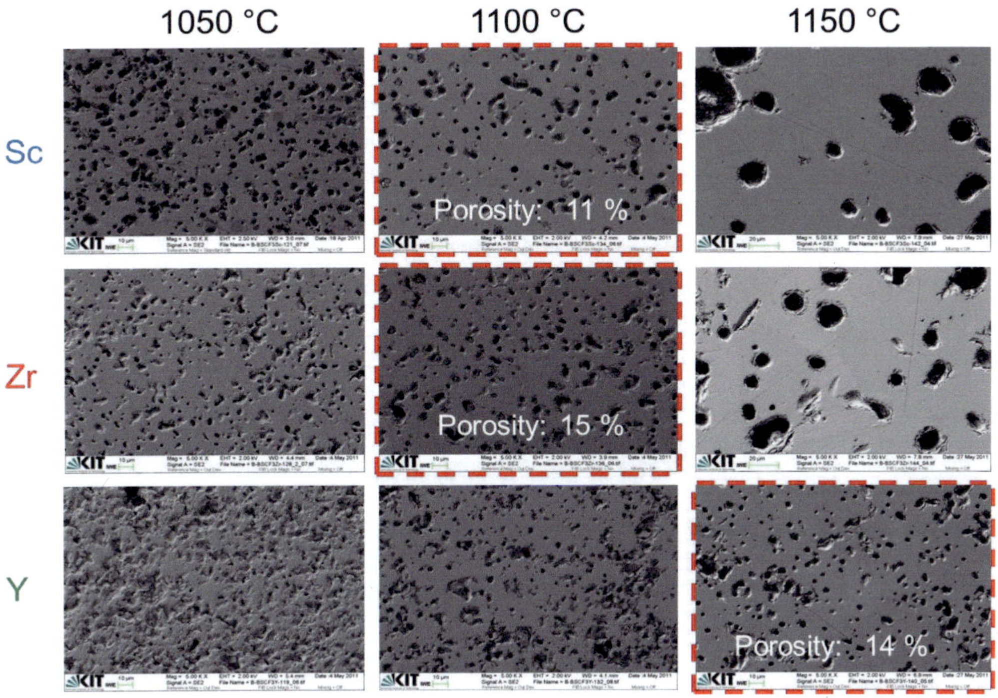

Fig. 2.10: SEM pictures of doped BSCF (3% Sc: top row, 3% Zr: middle row, 3% Y: bottom row) sintered for 12 h at different temperatures (1050 °C: left column, 1100 °C: middle column, 1150 °C: right column).

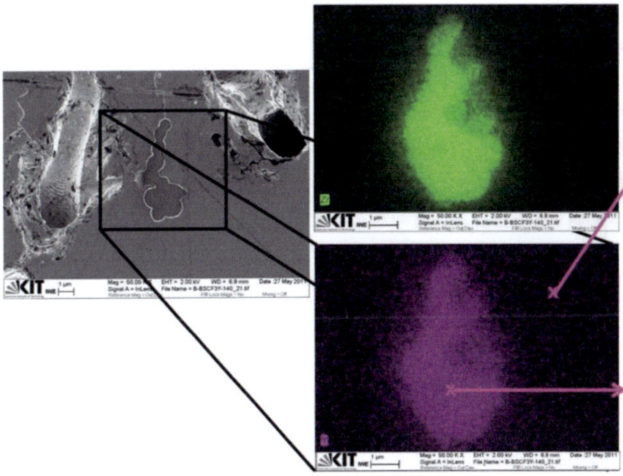

Element	Series	unn. C [wt.%]	norm. C [wt.%]	Atom. C [at.%]	Error [%]
Iron	K-series	6.14	7.22	11.00	0.2
Cobalt	K-series	23.52	27.66	39.95	0.6
Strontium	K-series	20.15	23.69	23.01	0.5
Barium	L-series	34.29	40.32	24.98	0.9
Yttrium	K-series	0.94	1.10	1.05	0.0
	Total:	85.03	100.00	100.00	

Element	Series	unn. C [wt.%]	norm. C [wt.%]	Atom. C [at.%]	Error [%]
Iron	K-series	2.11	2.51	3.99	0.1
Cobalt	K-series	7.02	8.37	12.58	0.2
Strontium	K-series	6.75	8.04	8.13	0.2
Yttrium	K-series	54.89	65.41	65.19	1.4
Barium	L-series	13.15	15.67	10.11	0.4
	Total:	83.92	100.00	100.00	

Fig. 2.11: SEM picture (left) and EDX mappings (middle) of sintered BSCF3Y and point analysis of secondary phase vs. bulk matrix (right).

While for electrical measurements on degradation phenomena in sintered samples simple 4-wire contacting with Pt wire and frit-free Pt paste fired at 1050 °C for 1 h in ambient air is totally sufficient (as the interest lies in relative changes in conductivity with temperature, time or pO_2), exact determination of the material's conductivity values as well as comparisons between different compositions require a precise electrical contact geometry. To achieve precisely determined inner contact distances, bulk samples used for these kinds of measurements have been contacted by magnetron sputtering (Von Ardenne CS730S) four lines of Pt onto the surface of the samples, which were subsequently contacted with Pt wire and paste, resulting in sharply defined inner contacts as shown exemplarily in Fig. 2.12.

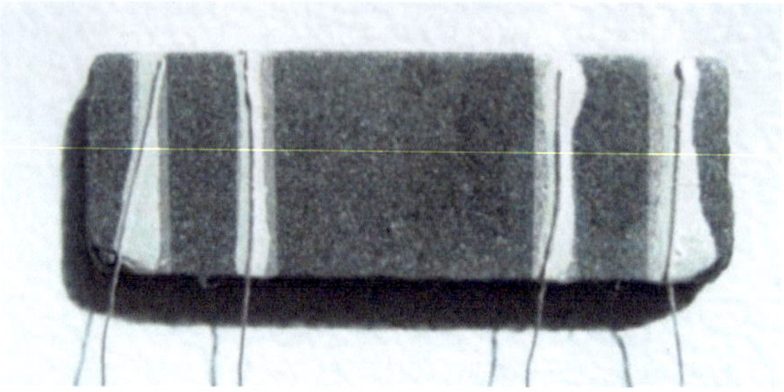

Fig. 2.12: Sputtered Pt-contacts on a sintered BSCF sample, contacted with sintered Pt paste and wires, resulting in a precisely defined inner contact distance.

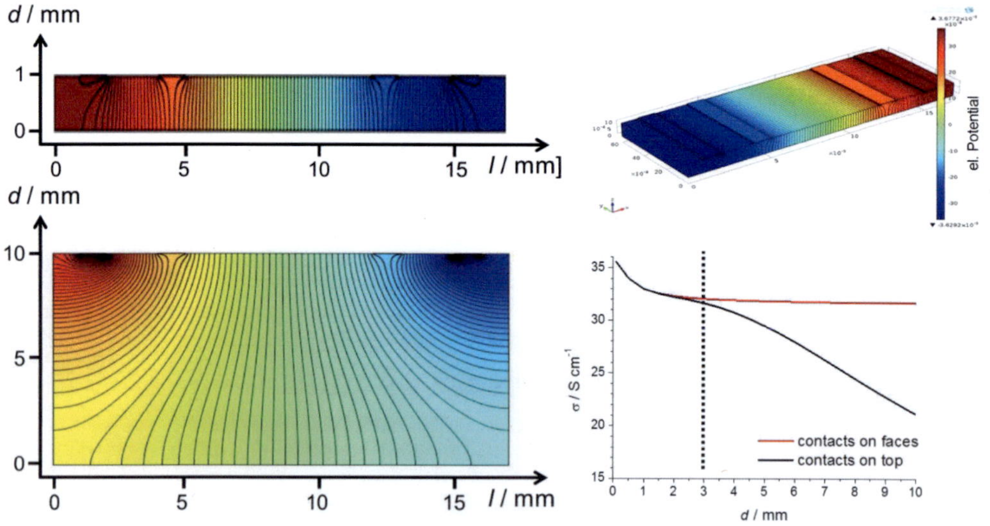

Fig. 2.13: Simulated equipotential planes in BSCF samples of different thicknesses. Current contacts on top are feasible up to $d = 2.5$ mm.

In order for this contacting method to be feasible, however, the samples must not exceed a certain thickness, over which the electrical field gradient in the sample volume becomes too distorted to accurately measure the intrinsic conductivity using only surface contacts for both the outer current contacts as well as the two inner measurement contacts. To this end, some basic simulations have been conducted in COMSOL Multiphysics, to visualize the deviation between all surface contacts and current contacts applied to the two faces of the sample (Fig. 2.13). Considering the conductivity of BSCF, pure surface contacting is feasible at thicknesses not exceeding 2.5 mm (bottom right in Fig. 2.13).

2.4.3 Epitaxial BSCF Thin Films

The preparation of the epitaxial BSCF thin films measured in this work has been performed by Mónica Burriel – during her stay at IWE – in cooperation with José Santiso at the Centro de Investigación en Nanociencia y Nanotecnologia, CIN2 (CSIC-ICN), Bellaterra/Spain and detailed in our paper [10].

Pulsed Laser Deposition (PLD) was used to grow epitaxial thin films of BSCF on $NdGaO_3$ (110) (NGO) single crystal substrates (dimensions 5 x 5 x 0.5 mm³, Crystal GmbH, Berlin, Germany). On these substrates, epitaxial growth was possible due to the only small mismatch with the basal plane of the BSCF perovskite structure, given that the orthorhombic NGO presents pseudo-cubic in-plane parameters of 3.863 and 3.854 Å, as compared to the 3.986 Å cubic lattice constant measured on the BSCF powder samples (cf. section 2.4.1. of this chapter).

Ceramic disc shaped PLD targets were prepared from BSCF powder produced at IWE (calcination temperature 1050 °C, 10 hours, with a D_{50} value after milling of 3.26 µm) by isostatic pressing (400 MPa, 5 min) and sintering at 1100 °C for 10 hours. These targets were single phase cubic BSCF according to XRD measurements, possessing the desired stoichiometry as verified by EDX.

A KrF excimer laser (λ = 248 nm) was used for the ablation, with a pulse rate of 10 Hz and an energy densitiy of 3 J/cm² per pulse. The number of pulses was varied from 4000 to 10000, while the substrate was kept at a temperature of 700 °C and at an oxygen pressure of 0.4 mbar. Before the actual film deposition, a pre-ablation was always performed prior to thin film deposition in order to ensure stoichiometric ablation.

Structural analysis measurements on the deposited layers were conducted at CIN2 using an X'Pert MRD PANalytical diffractometer. Fig. 2.14 shows the XRD pattern obtained for films of two different thicknesses *d*. The reflections from the NGO substrate are clearly visible and indicated with an asterisk. The only other reflections are 00*l* reflections of the BSCF phase, confirming the highly epitaxial growth of the layers.

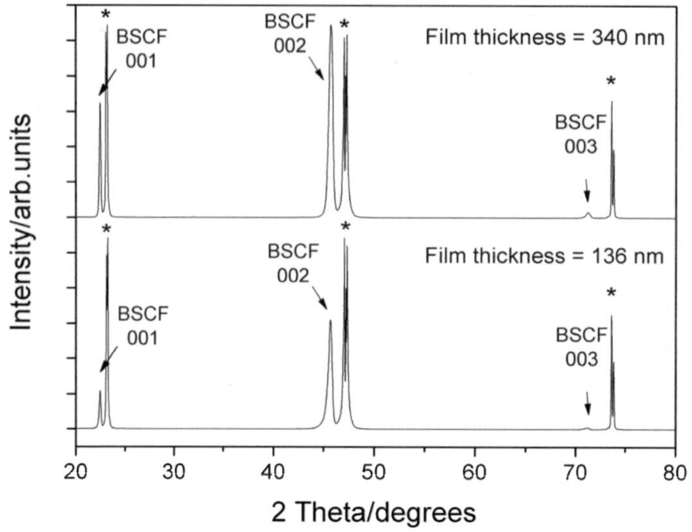

Fig. 2.14: XRD patterns of two BSCF films ($d = 340$ and 136 nm) on NGO. Substrate *hh*0 reflections are indicated with asterisks and BSCF film 00*l* reflections are labeled. From our joint publication [10].

The high degree of orientation in the perpendicular direction and the perfect in-plane alignment of the film have been confirmed by rocking curve and X-ray reflectometry measurements, as detailed in [10].

Fig. 2.15 shows surface SEM pictures of BSCF thin films with the two different thicknesses (136 nm (a) and 340 nm (b)). The layers appeared to be homogeneous, dense and without cracks, while the surface morphology turned out to be reasonably smooth, showing the visible crystal protrusions to be oriented in two main perpendicular directions. There was no delamination or reaction with the NGO substrate, as can be seen in the TEM images in Figs. 8.4 and 8.5 in Chapter 8.

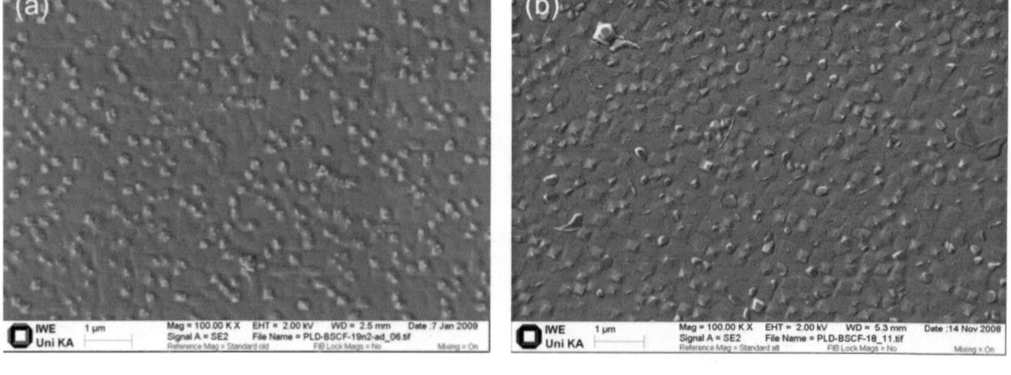

Fig. 2.15: SEM surface micrographs of a 136 nm (PLD-BSCF-19n2, (a)) and 340 nm thick (PLD-BSCF-18n1, (b)) BSCF thin film on NGO. From our joint publication [10].

All PLD thin films electrically measured in the following chapters of this work have been deposited with 10000 pulses and possessed a thickness of 340 nm (the PLD process allowed for four substrates to be coated simultaneously, therefore providing identical samples, resulting in excellent comparability).

These epitaxial thin films provide the opportunity to measure the intrinsic properties of BSCF without the influence of grain boundaries as well as – due to the very small thickness – purely surface controlled kinetics, but are somewhat influenced in their behavior due to stress and strain applied by the substrate, as will be discussed in the appropriate chapters. Additionally, due to the small sample volume, structural changes become easily visible, as shown in Fig. 2.16, where morphological changes become already visible after the firing of the electrical contacts.

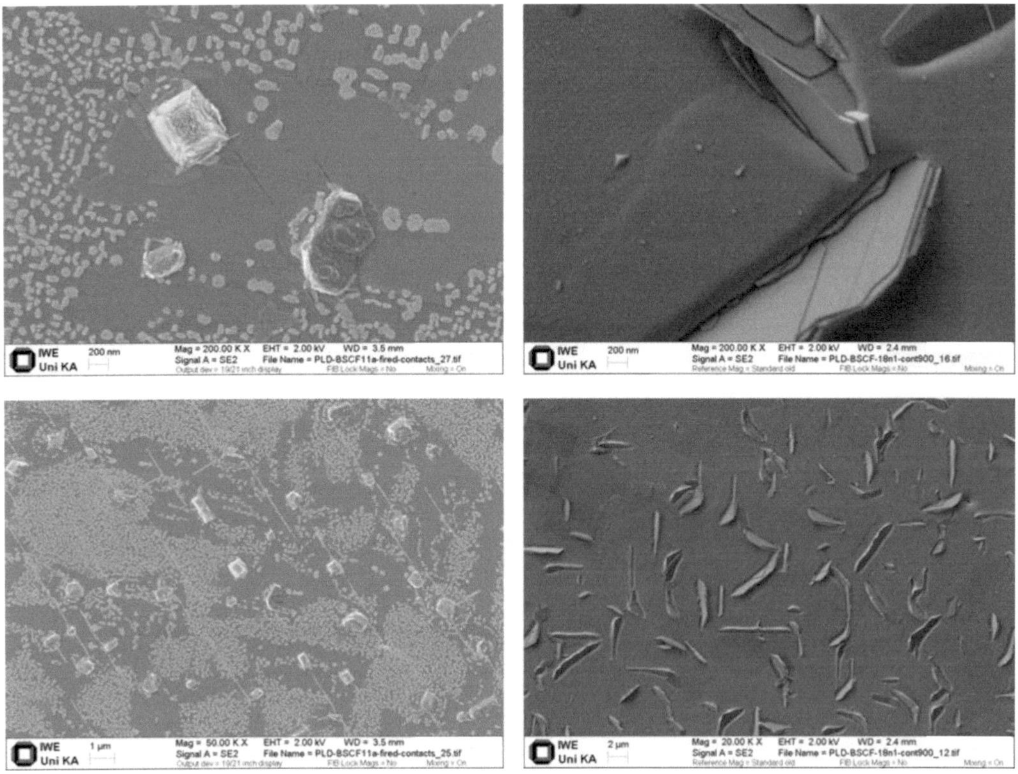

Fig. 2.16: SEM pictures of epitaxial thin films after contacting and burning contacts for 1 h at 1050 °C (left) and 950 °C (right).

The high temperature treatment for one hour at 1050 °C caused secondary phases to appear on the surface, probably similar to the condensed secondary phases inside pores in high-temperature sintered BSCF bulks (cf. Fig. 2.9 in section 2.4.2.) with enriched Sr-content as indicated by the EDX measurement shown in Fig. 8.5 in Chapter 8. During the treatment at 950 °C, plate like structures have formed in the BSCF layer, showing the thermal stability difficulties of the BSCF system discussed in Chapter 8. All PLD thin films have been contacted with two parallel wire contacts – either Au or Pt – using Au or Pt paste and firing for one hour at 950 °C or 1050 °C, respectively.

2.4.4 Polycrystalline BSCF Thin Films by Metal-Organic Deposition

The preparation of the polycrystalline BSCF thin films measured in this work has been performed by Koichi Asano from the Central Research Institute of Electric Power Industry (CRIEPI), Nagasaka, Yokosuka, Japan during his stay at IWE and is detailed in our paper [11].

A metal-organic deposition method (MOD) – in this case spin-coating – was used to produce polycrystalline BSCF thin-films on $NdGaO_3$ (110) (NGO) single crystal substrates (10 x 10 mm^2, thickness 0.5 mm; Crystec GmbH, Berlin, Germany). The substrate size was chosen four times larger than for the PLD layer preparation in order to minimize the influence of substrate edges on the spin coated layer. As for the PLD layers, NGO was chosen due to its chemical compatibility with BSCF as well as its very high electrical resistance at elevated temperatures, therefore not influencing any subsequent electrical measurements. The commercially available BSCF coating solution with the desired nominal composition was obtained from Kojundo Chemical Lab. Inc., Saitama, Japan.

Fig. 2.17: Schematic diagram of a metal organic deposition (MOD) method [11].

Fig. 2.17 shows a schematic of the two-step spin-coating process, consisting of a first step where the substrate was spin-coated with the solution using a rotation speed of 3000 rpm for 10 s and a second step at 4000 rpm for 24 s (a). The as-coated films were subsequently dried at 200 °C in air to evaporate the solvents (b). This procedure can be repeated several times, resulting in either one-time or multiple-time spin-coated thin films with increasing thickness.

For the following thermal annealing process (c), different annealing temperatures between 600 °C and 1000 °C as well as different temperature profiles were applied as described in our paper [11].

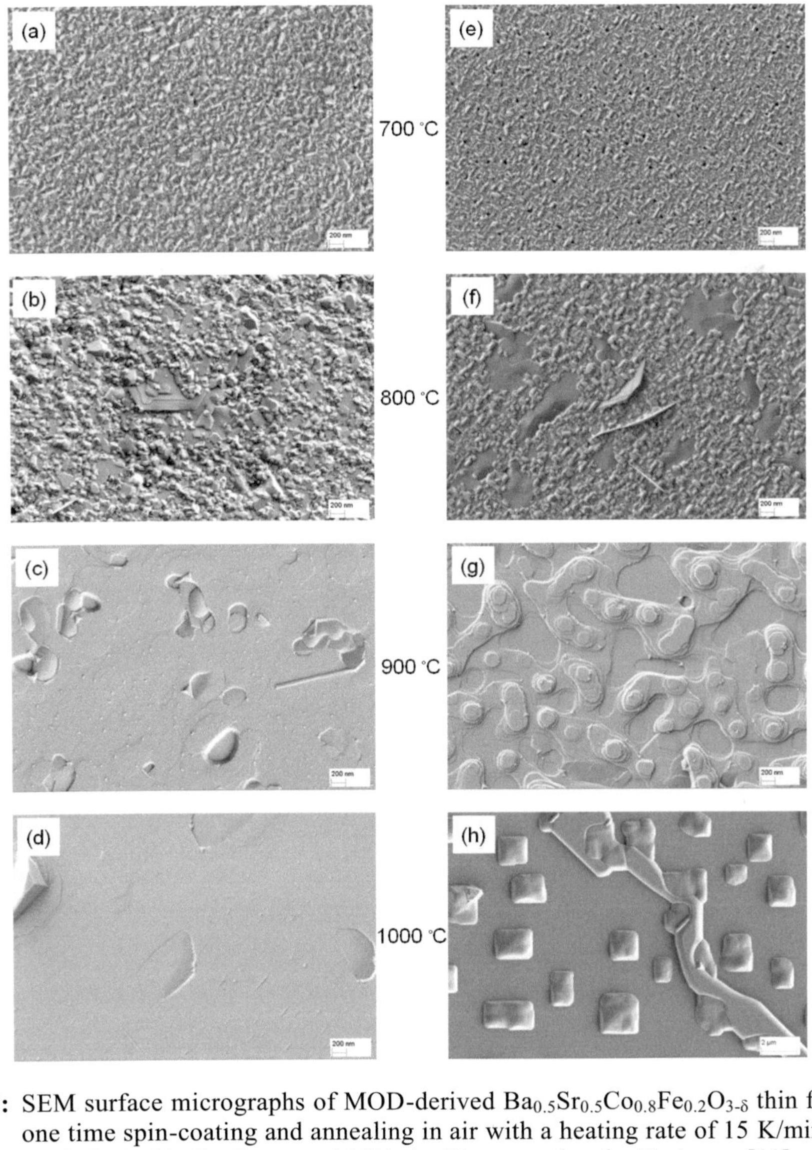

Fig. 2.18: SEM surface micrographs of MOD-derived $Ba_{0.5}Sr_{0.5}Co_{0.8}Fe_{0.2}O_{3-\delta}$ thin films after one time spin-coating and annealing in air with a heating rate of 15 K/min (a–d) or 3 K/min (e–h). Cooling rate 15 K/min. Pictures taken by K. Asano [11].

Fig. 2.18 shows SEM micrographs of the surface morphology variation of BSCF thin films on the NGO substrate after annealing at temperatures between 700 °C and 1000 °C, using heating rates of 15 K/min (a–d) and 3 K/min (e–h), and cooling rates of 15 K/min for all depositions. Depending on the annealing procedure, different surface morphologies could be obtained, e.g., oriented grain growth with a textured surface (e, f), terraced shapes (g), nano-sized particles (a,b) as well as relatively smooth, dense layers (c,d), with cubic BSCF forming predominantly at annealing temperatures of 900 °C and more.

Fig. 2.19 shows grazing-incidence XRD results obtained on a thin film sample after 5-time coating on a single crystal NGO substrate and subsequent annealing at 950 °C with heating and cooling rates of 15 K/min (top left). In order to subtract the reflections of the substrate, uncoated NGO was measured as a reference (top right) and used to correct the data to obtain the spectrum generated by the BSCF layer. Bottom right in Fig. 2.19 is the result (red curve) in comparison with the spectrum obtained from cubic BSCF powder (blue curve). No background has been subtracted in all shown curves, except for the BSCF powder reference.

The achievable BSCF film thickness on NGO substrates using this temperature, annealing profile and 5-times coating was between 50 and 80 nm, while exhibiting the smoothest surface morphology of all samples shown in Fig. 2.18 and [11].

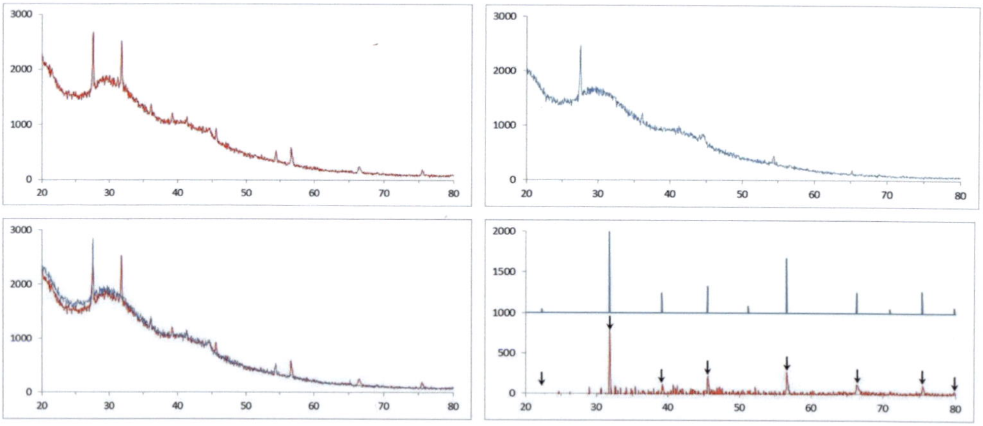

Fig. 2.19: Grazing incidence XRD measurements on a BSCF MOD film on NGO (top left), the NGO substrate itself (top right), an overlay of both measurements (bottom left) and the differential (red curve on bottom right).

Fig. 2.20 shows a SEM micrograph of the surface morphology of this sample. It displays a smooth, slightly terraced surface with some nano-sized particles, presumably representing the same secondary phase as detected on the PLD thin films (cf. section 2.4.3 and Chapter 8).

While this investigation of the different obtainable morphologies opens possibilities for detailed surface tailoring to meet different requirements for several applications (e.g. surface activation for OTMs), all samples measured in this work have been investigated to obtain information on the intrinsic properties of the BSCF system and therefore the smoothest samples – all prepared under the stated conditions – have been used throughout this work.

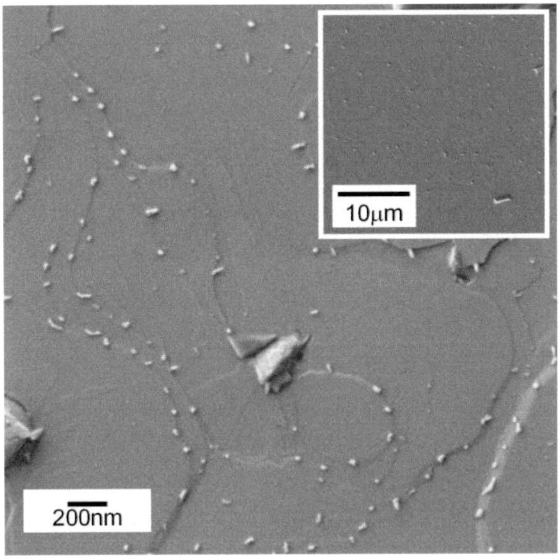

Fig. 2.20: SEM surface image of BSCF thin film with 5 times spin-coating on NGO single crystal substrate, RTA annealing at 950 °C with heating/cooling rates of 15 K/min. Picture taken by K. Asano [11].

References

[1] K. Beetz, *Die geschlossene Festelektrolyt-Sauerstoffpumpe*, Düsseldorf: VDI Verlag GmbH (1993).

[2] C. Niedrig, W. Menesklou, S. F. Wagner and E. Ivers-Tiffée, "High-Temperature pO_2 Stability of Metal Oxides Determined by Amperometric Oxygen Titration", *Journal of The Electrochemical Society* **160**, pp. F135-F140 (2013).

[3] C. Niedrig, S. F. Wagner, W. Menesklou and E. Ivers-Tiffée, "Characterization of oxygen-dependent stability of selected mixed-conducting perovskite oxides", *Solid State Ionics* **273**, pp. 41-45 (2015).

[4] C. Tragut and K. H. Härdtl, "Kinetic Behaviour of Resistive Oxygen Sensors", *Sensors and Actuators B-Chemical* **4**, pp. 425-429 (1991).

[5] S. F. Wagner, W. Menesklou, T. Schneider and E. Ivers-Tiffée, "Kinetics of oxygen incorporation into $SrTiO_3$ investigated by frequency-domain analysis", *Journal of Electroceramics* **13**, pp. 645-651 (2004).

[6] S. Švarcová, K. Wiik, J. Tolchard, H. J. M. Bouwmeester and T. Grande, "Structural instability of cubic perovskite $Ba_xSr_{1-x}Co_{1-y}Fe_yO_{3-\delta}$", *Solid State Ionics* **178**, pp. 1787-1791 (2008).

[7] H. Koster and F. H. B. Mertins, "Powder diffraction of the cubic perovskite $Ba_{0.5}Sr_{0.5}Co_{0.8}Fe_{0.2}O_{3-\delta}$", *Powder Diffraction* **18**, pp. 56-59 (2003).

[8] T. Schneider, *Strontiumtitanferrit-Abgassensoren: Stabilitätsgrenzen/Betriebsfelder*, Aachen: Verlag Mainz (2005).

[9] H. Wang, W. Yang, Y. Cong, X. Zhu and Y. S. Lin, "Structure and oxygen permeability of a dual-phase membrane", *Journal of Membrane Science* **224**, pp. 107-115 (2003).

[10] M. Burriel, C. Niedrig, W. Menesklou, S. F. Wagner, J. Santiso and E. Ivers-Tiffée, "BSCF epitaxial thin films: Electrical transport and oxygen surface exchange", *Solid State Ionics* **181**, pp. 602-608 (2010).

[11] K. Asano, C. Niedrig, W. Menesklou, S. F. Wagner and E. Ivers-Tiffée, "Modification of Oxygen / $(Ba_{0.5}Sr_{0.5})(Co_{0.8}Fe_{0.2})O_{3-\delta}$ Interfaces Derived By Metal-Organic Deposition", *ECS Transactions* **66**, pp. 147-155 (2015).

Chapter 3

Coulometric Titration in an Oxygen Pump Setup

3.1 Introduction

An assessment of basic material properties should also include knowledge of the chemical stability of the MIEC oxide as a function of temperature as well as ambient oxygen partial pressure pO_2. However, this stability range is not always easily determined. In the case of BSCF for instance, there have been several studies on its thermal stability, e.g. [5-9], see also Chapter 8. Even though there have also been several reports on its oxygen non-stoichiometry in the past, e.g. [10-12], many statements in literature have only given a very rough estimate of its stability limits at lower pO_2, e.g., [13;14], presumably due to experimental limitations. Apart from our previously published study [4], another investigation on the low-pO_2 stability limits of BSCF has only been reported recently [15], followed by parts of the investigations shown in this chapter and published in our paper [2].

A common method for pO_2-dependent stability investigations makes use of setting a defined pO_2 value with the help of gas mixtures. However, such a setup suffers from lack of *continuous* adjustability of pO_2 down to very low values. In this thesis, therefore, a custom-made closed tubular zirconia "oxygen pump" setup is employed (cf. Fig. 3.1 and Section 2.3.2 in Chapter 2). This setup was originally devised by Beetz [16] and has since been adapted and improved with respect to titration experiments. An YSZ (ZrO_2 doped with Y_2O_3) tube with pasted platinum electrodes facilitates precise control – via the Nernst equation (cf. eq. 2.1 in Chapter 2) – of the oxygen partial pressure $pO_{2,inside}$ in its enclosed volume within the whole pO_2 range between 10^{-18}...1 bar at temperatures above approximately 700 °C.

Parts of the results shown in this chapter have been published in [1-4].

By measuring the electric current necessary for pumping oxygen through the solid zirconia electrolyte into or out of the sample chamber in order to obtain the desired pO_2 set value, the amount of oxygen transported can be determined with a high resolution. Each oxygen molecule transported through the tube (as O^{2-} ions) requires an external counter-transport of four electrons. It is thus possible to determine the oxygen uptake or release of an oxide sample as a function of the controlled ambient $pO_{2,inside}$, temperature, and measurement time with high precision.

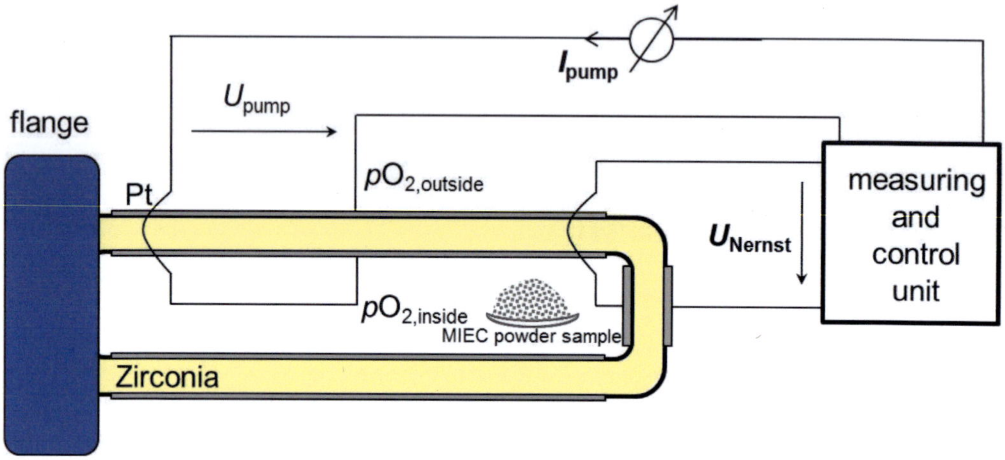

Fig. 3.1: Schematic of the “oxygen pump” measurement setup with a powder sample.

Similar titration methods with solid-electrolyte cells have been employed by various research groups in the past for coulometric or amperometric studies of the oxygen non-stoichiometry and, ultimately, the chemical stability of high-temperature solid oxide materials as a function of pO_2, cf., e.g. [17-20]. In most cases, however, an experimentally rather laborious setup is involved, in some cases not even facilitating non-destructive testing. Sealing at temperatures as high as 900 °C also is a mostly very intricate issue to be dealt with.

The coulometric titration setup presented in this thesis provides several advantages with respect to experimental simplicity. The setup can be re-used many times, a very convenient exchange of samples is possible, and the sealing of the apparatus is achieved without any difficulty at room temperature. This, on the other hand, requires larger dimensions and, hence, a fairly large dead volume so that each titration measurement of a sample must be corrected with an additional calibration (background) measurement of the empty setup.

However, it has been experienced that several parasitic effects, such as pO_2-dependent contributions of the zirconia tube itself to the overall pump current, possible leakage currents etc., have to be taken into account, especially under very reducing conditions, in order to avoid measurement errors. Therefore, the background measurement has to be carried out anyway. These issues are addressed in Section 3.3, as well as advantages and limitations of the method.

In order to validate the setup for the coulometric titration method presented in this thesis, a simple binary "model" oxide (copper oxide) is used, where pO_2-dependent transition conditions are known from phase diagrams in literature, e.g. [21]. To verify the oxygen partial pressure measured in the setup, an electrical measurement was carried out on an $SrTiO_3$ single crystal. The defect chemistry of this model perovskite has been extensively studied over the past decades, which can therefore serve as a reliable oxygen-sensing standard.

After the successful quantitative validation, the method is then applicable to more complex (and application-oriented) MIEC materials, such as LSCF and BSCF. It is shown that the coulometric oxygen titration method is able to yield information on oxygen stoichiometry changes during continuous pO_2 changes and, finally at low-pO_2 values, irreversible demixing phenomena. Thus, the stability limits of the MIEC material can be easily determined.

3.2 Experimental

3.2.1 Powders

Commercially available copper oxide powder (CuO) was obtained from Riedel-de Haën (Germany); XRD and particle-sizer measurements revealed single-phase CuO with a monomodal particle size distribution (D_{50} value of 0.3 µm).

Titration and annealing experiments were carried out with the different powders (cf. Chapter 2) deposited into an open Pt crucible.

Different powder samples annealed at 900 °C for 10 h in reducing atmospheres (cf. XRD spectra in Fig. 3.9) have been cooled down with the maximum cooling rate possible in the oxygen pump setup (1 K/min) while maintaining the preset pO_2 as long as oxygen-ion conduction through the YSZ tube is still possible.

3.2.2 Electrically Contacted $SrTiO_3$ Single Crystal Sample

For conductivity measurements a nominally undoped $SrTiO_3$ single crystal (Crystec, Berlin/Germany, $5 \cdot 5 \cdot 0.04$ mm³) was contacted with Pt wires using a frit-free Pt paste fired at 1050 °C in ambient air. Owing to the fairly high electrical resistance, electrical measurements could be carried out in 2-point technique.

3.3 Results and Discussion

3.3.1 Intrinsic Properties of the Oxygen-Pump Setup

Assuming that oxygen can only be transported into or out of the sample chamber by oxygen ion transport through the YSZ, monitoring of the current I_{pump} yields information about the amount of transported oxygen: By integrating I_{pump} over time t, the transported charge and therefore the number of oxygen molecules N_{O_2} can be calculated (e being the elementary charge):

$$N_{O_2} = \frac{1}{4e} \int I_{pump} \mathrm{d}t \tag{3.1}$$

The measured current I_{pump} during a titration experiment has to be corrected for any intrinsic setup properties. Extensive background measurements have shown that there are multiple influences on the pump current. First, there is a non-negligible constant offset (between 200...700 µA) present in all background measurements that is not depending on pO_2, but rather on the specifically used – exchangeable – pump tube (and its age, i.e. its accumulated overall operating time) as well as electronic properties of the controller itself. It can be shown that the onset of electronic conductivity in YSZ at low oxygen partial pressures does not by far amount to these currents: At an oxygen partial pressure of 10^{-4} bar (corresponding to an applied Nernst voltage drop of 232.8 mV at 900 °C across the tube wall thickness of 2 mm, if measured against pure oxygen), this leads to an electronic current of only around 0.2 µA, according to Park and Blumenthal [22].

Nevertheless, Fig. 3.2 shows background measurements displaying increasing currents below pressures of approx. 10^{-10} bar. These can also not be explained by the possible presence and subsequent dissociation of H_2O molecules, as the characteristics proved to follow no trend with respect to H_2O vapor pressure inside the sample space. Further in-depth analysis (not shown here) revealed that this additional transported charge can be associated with oxygen being released by the YSZ tube itself at applied voltages above ~0.8 V in these low-pO_2 atmospheres in a partially reversible process, thereby also contributing to the aging of the pump tube. This oxygen release already at Nernst potentials below 1 V has also been reported by Werner and Schmid-Fetzer [23] for 8YSZ and 5YSZ as well as by Shah et al. [24] for ceria-zirconia solutions, most notably $Ce_{0.25}Zr_{0.75}O_{2-x}$. Yet, this oxygen release is not detrimental to achieving exact and even lower partial pressures at the sample position, as can be monitored by the Nernst voltage between the reference electrodes in direct vicinity.

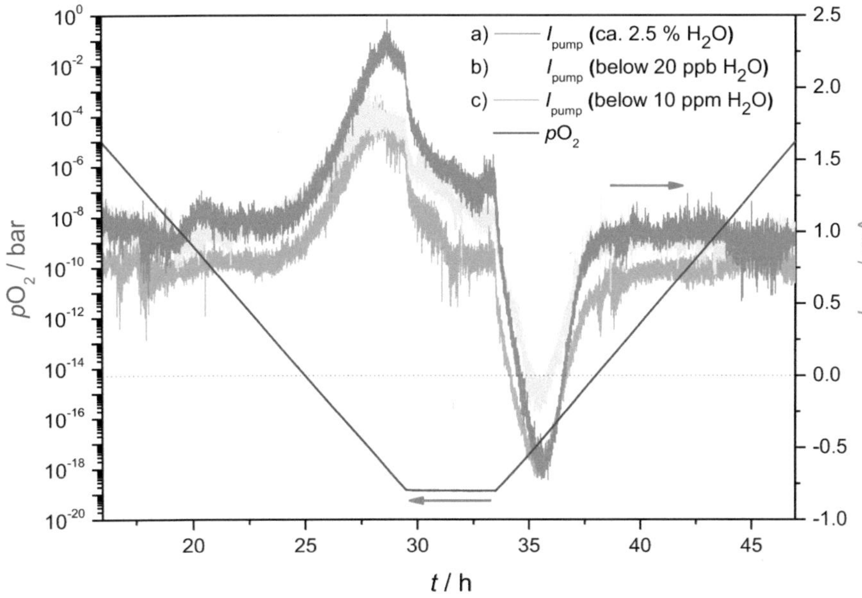

Fig. 3.2: Background measurements at different water contents. The water content of the oxygen used for flushing the sample chamber was varied by using a) pure oxygen (Air Liquide, Germany, 99.95 % purity) in combination with a gas-washing bottle (providing a humidification of $xH_2O \approx 0.025$ [16]), b) pure oxygen with a H_2O filter cartridge (Air Liquide Hydrosorb, $xH_2O \leq 20$ ppb), and c) pure oxygen from the bottle (Air Liquide, Germany, water fraction: $xH_2O \leq 10$ ppm). The current characteristics show no significant dependence on H_2O vapor pressure inside the sample space.

However, the process obviously necessitates the observed adjustments in the pump current, which additionally implies that while at higher pO_2 a correction of the measured pump current is possible by simply subtracting the setup-dependent offset, all measurements below 10^{-10} bar pO_2 have to be corrected by performing a mandatory previous and/or subsequent background measurement with the exact same pO_2-profile.

3.3.2 Validation of Titration Method

Feasibility of this coulometric titration method can be nicely demonstrated using a simple binary "model" oxide where chemical stability and pO_2-dependent transition conditions are known from phase diagrams in literature. A comparison of transition conditions for different Oxides with the relevant temperature and pO_2 range for the oxygen pump setup is shown in Fig. 3.3. One suitable material is CuO where according to Neumann et al. [21] reduction to Cu_2O should occur (at $T = 850$ °C) at $pO_2 \approx 10^{-2}$ bar.

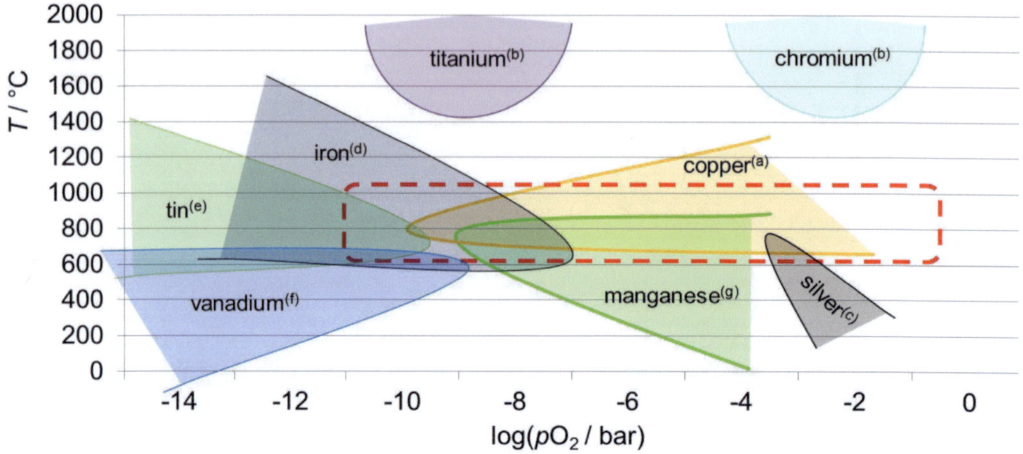

Fig. 3.3: Decomposition ranges of different metal oxides in comparison to relevant measurement ranges in the oxygen pump setup. From: (a) [21], (b) [25], (c) [26], (d) [27], (e) [28], (f) [29], (g) [30].

In Fig. 3.4, the coulometric titration curve of CuO powder is shown in this pO_2 range, as compared to the background measurement of the empty pump cell under identical conditions. The huge increase in pump current at $pO_2 \approx 4 \cdot 10^{-3}$ bar reflects this reduction process. The large amount of oxygen set free from the CuO (half its oxygen is released as gaseous O_2 until only Cu_2O remains) leads to a buffering effect:

The pO_2 within the chamber stays at a fairly constant value for several hours until the reduction is completely finished. This is nicely visible at $pO_2 \approx 5 \cdot 10^{-3}$ bar, where the partial pressure in the sample space stays almost constant over several hours while the pump current increases to the physical maximum of the pump setup. The well-defined drop in the current and the nearly instantaneous adjustment to the preset final pO_2 value (background measurement, cf. cyan dashed curve in Fig. 3.4) strongly suggest that CuO has now completely been reduced to Cu_2O.

By integrating the measured $I_{pump}(t)$ (cf. green dashed boundaries in Fig. 3.4, corrected by I_{pump} of the empty setup) a charge of 531 C can be determined. Taking into account that 465 mg CuO were titrated, complete reduction would result in a theoretical release of $8.8 \cdot 10^{20}$ oxygen molecules. Their complete removal from the chamber via the zirconia tube would require an electronic charge transport of 564 C. This is in very good agreement (only 6 % deviation) with the value obtained experimentally.

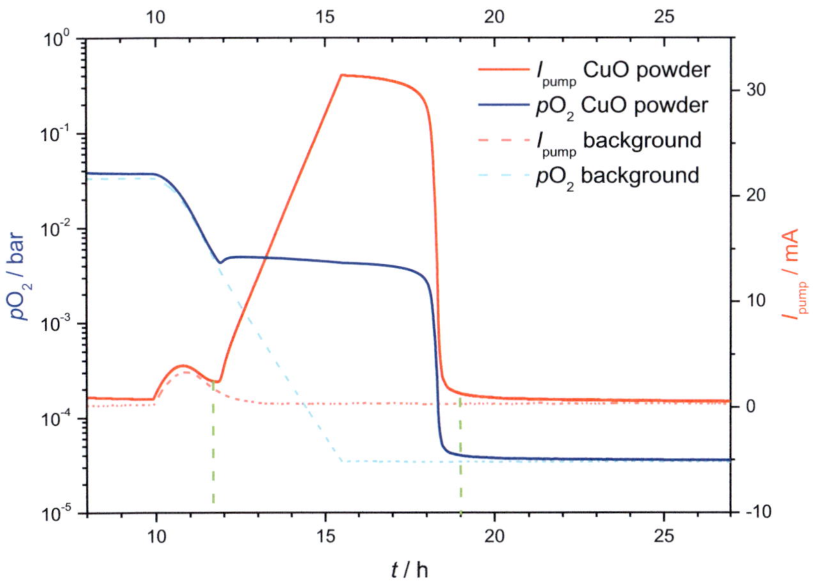

Fig. 3.4: Coulometric titration of CuO powder (red and blue curves) at $T = 850$ °C between $pO_2 \sim 0.03$ and $\sim 3 \cdot 10^{-5}$ bar, in comparison to a background measurement of the empty setup (no powder sample: light red and cyan dashed lines) with identical parameters. Green vertical lines indicate integration boundaries.

Additional measurements at 800 °C and 900 °C yielded partial pressure values for the reduction and subsequent reoxidation of CuO and Cu_2O, respectively. These values are in very good agreement with the phase diagram in Fig 3.5.

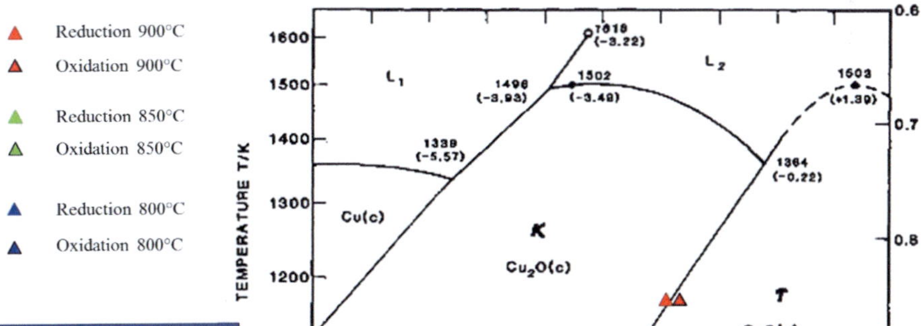

T / K	Reduction $\log(pO_2)$	Oxidation $\log(pO_2)$
1073	-3,296	-2,573
1123	-2,431	-2,154
1173	-1,816	-1,661

Fig. 3.5: Reduction and oxidation conditions measured for copper oxide in comparison to the phase diagram [21].

3.3.3 Verification of the Oxygen Partial Pressure in the Setup

Regarding the question concerning the accuracy of the oxygen partial pressure measurement (reference sensor close to the sample position in Fig. 3.1), a control measurement was carried out using an electrically contacted nominally undoped strontium titanate single crystal. The defect chemistry of the perovskite oxide $SrTiO_3$ has been extensively studied over the past decades, its defect model is therefore well-known [31]. Using this defect model it was possible to fit the measured $\sigma(pO_2)$ values and, thus, to confirm the accuracy of the preset pO_2 values at the sample position within the measurement setup. As Fig. 3.6 shows, the preset pO_2 values could be confirmed between 10^{-17}...0.1 bar and 750...900 °C with an excellent accuracy. The fitted curves have been calculated with an intrinsic acceptor concentration of $N_A = 1.5 \cdot 10^{18}$ cm^{-3} using the well-established defect model by Moos and Härdtl [31]. The excellent agreement between fitted and measured curves shows that the preset pO_2 values in the "oxygen pump" setup indeed correspond to the real pO_2 values at the sample position throughout the measured pO_2 range 10^{-17}...0.1 bar (note that the defect model neglects ionic conductivity contributions; therefore, the measured conductivity minima deviate slightly from the model).

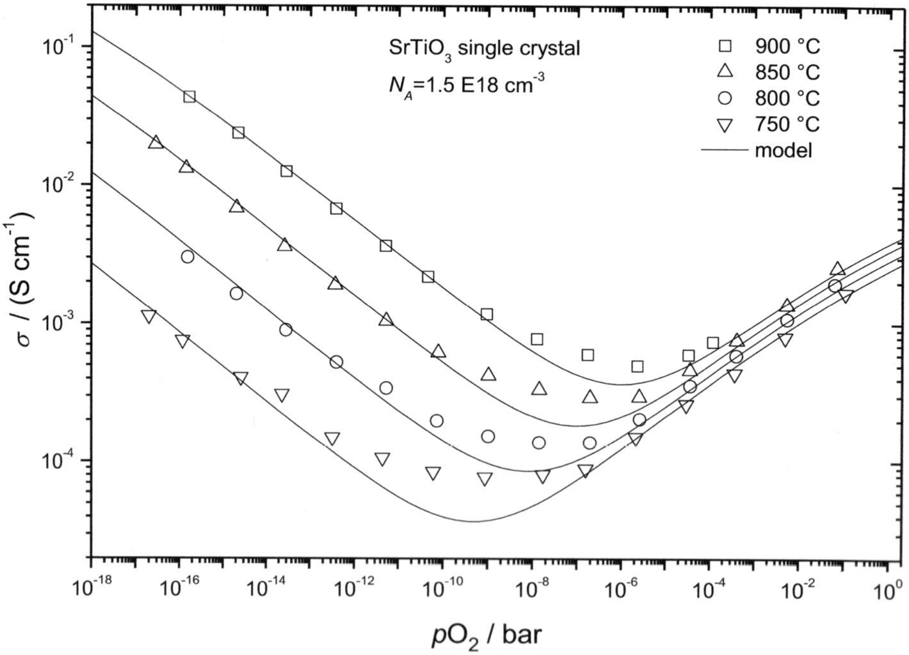

Fig. 3.6: Electrical conductivity of a strontium titanate single crystal measured as a function of oxygen partial pressure pO_2 in the oxygen pump setup at temperatures in the range of $750 < T$ / °C < 900.

3.3.4 Application to a Complex MIEC Oxide

After this successful quantitative proof of concept, the method is now applied to BSCF. Here, information with regard to exact stoichiometry changes, in particular with respect to low-pO_2 chemical stability due to irreversible demixing phenomena, has been published only very recently by Mueller et al. [15].

Following the line of thought from the last section one can deduce that the pump current necessary to pump the gaseous O_2 out of the zirconia tube is a very sensitive means of detecting the stability limits of the oxide: At a certain pO_2 value, a significant increase in the necessary pump current clearly indicates the beginning decomposition of the powder sample with its very large free surface area.

However, in such a complex materials system several oxygen-exchange related processes are to be expected. For a quantification of the oxygen stoichiometry in the cubic BSCF system it is paramount to be able to distinguish between these different processes. It is therefore necessary to take into account both the amount of oxide powder sample and the time-scale of the continuous pO_2 change. The amount of substance must be large enough to ensure a good signal-to-noise ratio (SNR) but small enough that the gaseous oxygen resulting from the decomposition processes can still be pumped out of the setup within a reasonable time.

The time-scale of the continuous pO_2 change must correspond to the oxygen-exchange and decomposition kinetics of the sample. Fig. 3.7 illustrates this issue:

Titration experiments were carried out at 900 °C (in order to avoid formation of the hexagonal phase, cf. Chapter 8 and [8]) on equal amounts of BSCF powder at different rates of pO_2 change (dU_{pump}/dt = 3 mV/min, 1 mV/min, and 0.1 mV/min, respectively). The fastest pO_2 change (top curve in Fig. 3.7) results in a very large amount of oxygen quickly set free. This amount cannot be pumped out of the setup fast enough leading to an increasing discrepancy between set and real pO_2 value during the measurement. The run of the measured pO_2 curve and the very slight bumps in the pump current curve lead one to suspect that several different pO_2-dependent processes occur which cannot be separated. As a result, a quantification of the pump current in order to assess the oxygen non-stoichiometry of the sample clearly is not possible.

Decreasing the pO_2 change rate by a factor of 3 (middle curve in Fig. 3.7) leads to a separation of the pO_2-dependent oxygen-exchange/decomposition processes (clearly visible as separate peaks in the pump current as well as the observed buffering effect in pO_2); however, the run of the measured and the preset pO_2 curves still show occasional deviations.

A further decrease of pO_2 change rate by a factor of 10 (bottom curve in Fig. 3.7) finally facilitates a very clear separation of the individual oxygen-exchange/decomposition processes, and the run of the measured and preset pO_2 curves shows that the system is very close to equilibrium throughout the measurement. (One must bear in mind, however, that such a slow measurement exceeds a reasonable time-scale, adds to the aging of the zirconia tube, and results in a decrease in signal quality, visible by the bad signal-to-noise ratio of the current measurement).

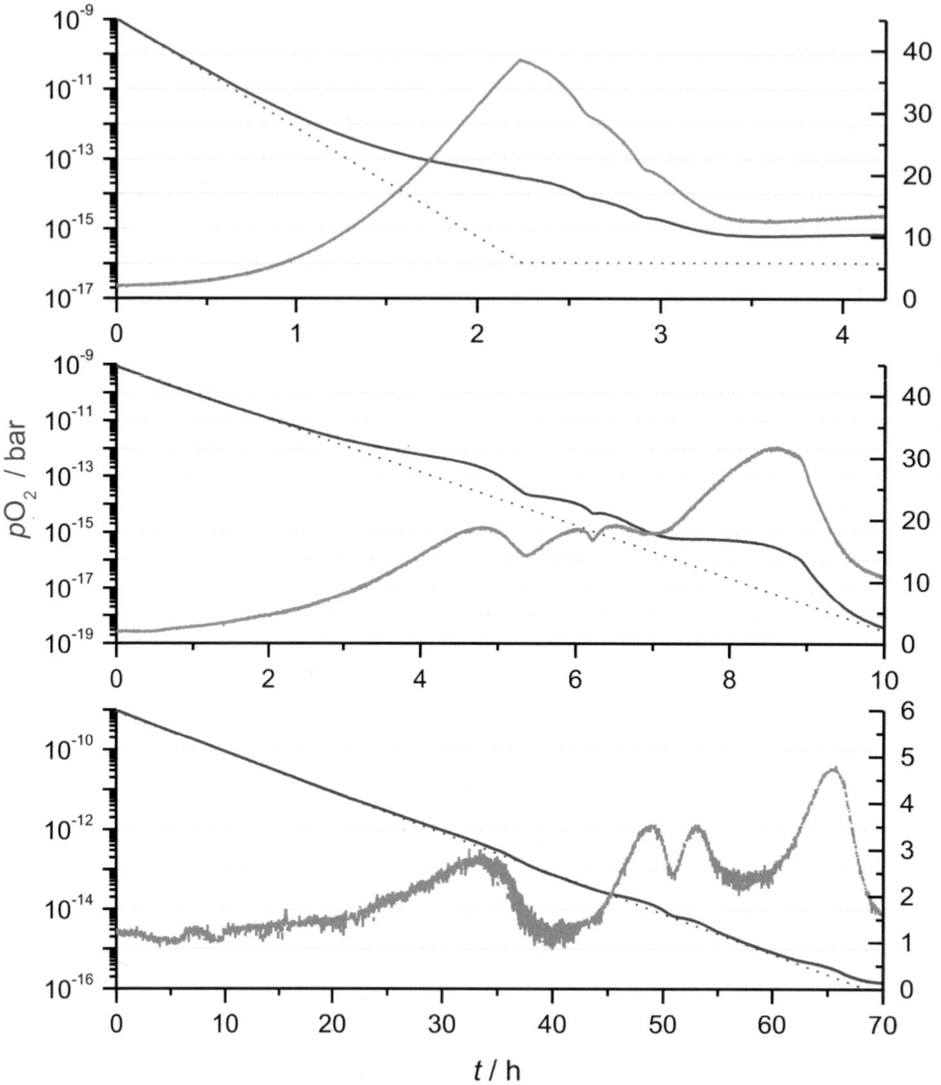

Fig. 3.7: Coulometric titration curves carried out on equal amounts of unused $Ba_{0.5}Sr_{0.5}Co_{0.8}Fe_{0.2}O_{3-\delta}$ (BSCF) powder at different rates of pO_2 change (dU_{pump}/dt = 3 mV/min (top), 1 mV/min (middle), and 0.1 mV/min (bottom), respectively). The red curves show the pump current (right axes), whereas the blue curves depict the actual pO_2 at the sample position. The dotted blue curves show the preset pO_2 ramps.

Fig. 3.8 shows the coulometric titration curve at T = 900 °C (dU_{pump}/dt = 1 mV/min) for the pO_2 range below 10^{-6} bar as well as the corresponding background measurement. By monitoring the pump current, several processes are observed (in reproducible experiments) where oxygen is set free from the sample (around 10^{-12}, 10^{-14}, 10^{-15}, and 10^{-16} bar).

One can, in principle, clearly distinguish two regimes: At pO_2 values above ca. $5 \cdot 10^{-13}$ bar the BSCF reversibly excorporates oxygen, resulting in increasing δ values (in the measurement, the deviation in pO_2 within the sample chamber resulting from these oxygen non-stoichiometry changes becomes clearly visible below $\sim 10^{-10}$ bar). At pO_2 values of around $5 \cdot 10^{-13}$ bar, the maximum of the first current peak is reached; this supposedly heralds the lower stability limit.

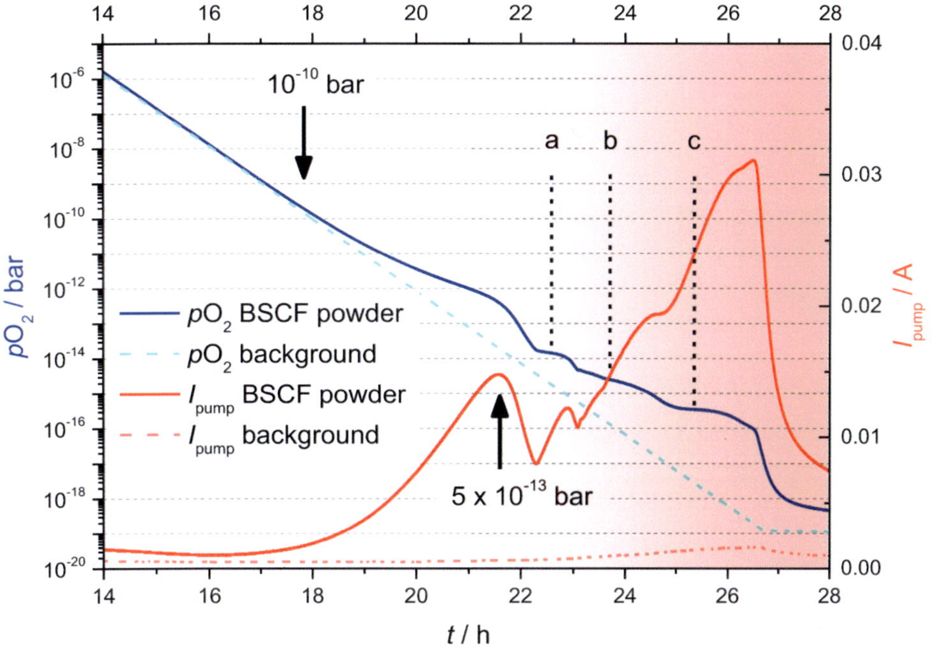

Fig. 3.8: Coulometric titration ($dU_{pump}/dt = 1$ mV/min) of cubic-phase BSCF powder at $T = 900$ °C and low partial pressures (red and blue curves), in comparison to a background measurement of the empty setup (light red and cyan dashed lines) with identical parameters.

At even lower pressures (shaded pO_2 range), BSCF is no longer chemically stable, cf. XRD measurements shown in Fig. 3.9. Several pO_2-dependent decomposition processes (a, b, and c) are visible which lead to peaks in the pump current and determine the overall pO_2 in the chamber until each process is finished. The decomposition kinetics appear to be fairly fast, as compared to various other MIEC perovskites characterized (cf. Section 3.3.6).

This is verified by XRD measurements on powders annealed at 900 °C and various oxygen partial pressures (for an annealing time of $\Delta t = 10$ h each), where after annealing at 10^{-13} bar pO_2 (or lower), XRD measurements clearly show that BSCF has decomposed (secondary-phase reflections, cf. Fig. 3.9).

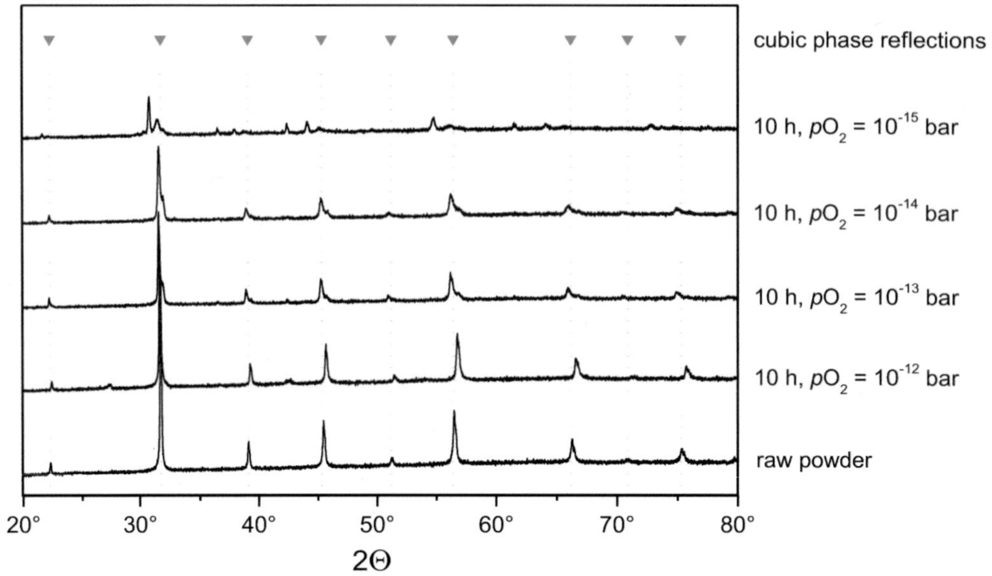

Fig. 3.9: Ex-situ X-ray diffraction (XRD) spectra of (originally single-phase cubic) BSCF powders annealed at 900 °C and various oxygen partial pressures (for $\Delta t = 10$ h each). No background subtraction has been performed in order to guarantee comparability of the measurements.

By employing this coulometric titration method to BSCF powder samples and in combination with an XRD analysis, it could be shown in [4] that the chemical stability limits of BSCF at temperatures between 700 °C and 900 °C are considerably lower than previously assumed in literature. Amongst the decomposition products, CoO and a remaining cubic phase stood out prominently (cf. Section 3.3.5).

Lastly, the oxygen stoichiometry changes could also be estimated using this method. After equilibrating 501 mg of BSCF powder at 10^{-3} bar and subsequent titration, the pump current was integrated with respect to time, starting from 10^{-4} bar until the minimum after the second current peak in Fig. 3.8 was reached ($\sim 5 \cdot 10^{-15}$ bar). The first peak corresponds to an actual pO_2 value of approx. $5 \cdot 10^{-13}$ bar, marking the stability limits of BSCF.

Fig. 3.10 shows the oxygen non-stoichiometry changes ($\Delta\delta$) for BSCF at 900 °C and partial pressures between $\sim 5 \cdot 10^{-15}$ and 10^{-4} bar. A sudden decrease in $\Delta\delta$ occurring at partial pressures of 10^{-12} bar and below indicates the end of the stability range for the cubic BSCF phase and confirms the decomposition limit previously determined by XRD measurements [4]. Below this value, the determination of $\Delta\delta$ becomes no longer meaningful but is shown for the sake of visualizing the end of the first and the beginning of the second decomposition process (cf. Fig 3.8).

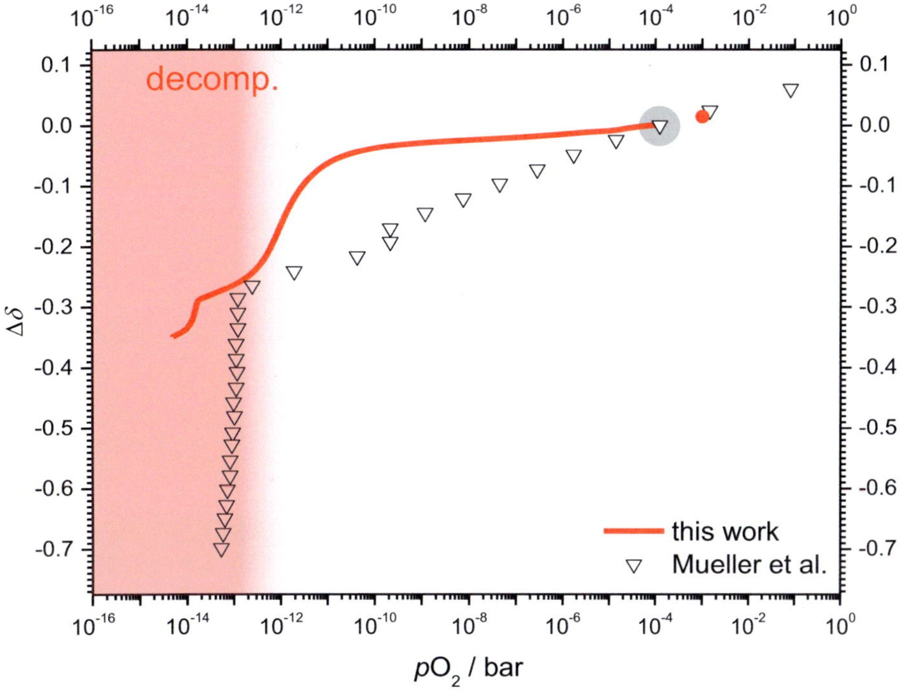

Fig. 3.10: Values for oxygen non-stoichiometry $\Delta\delta$ in BSCF as $f(pO_2)$ at 900 °C. Data from Mueller et al. [15] are shown for comparison.

Non-stoichiometry data of BSCF very recently published by Mueller et al. [15] is shown for comparison; as an "initial value $\Delta\delta = 0$" the value for $pO_2 = 10^{-4}$ bar was used (grey circle) because a quantification of titration experiments at higher pO_2 is quite difficult due to the disadvantageous ratio of gaseous oxygen from the dead volume and oxygen released from the sample. However, in preliminary measurements in the pO_2 region above 10^{-4} bar the slope of the $\Delta\delta$ curve appears to be similar to the data of Mueller et al. [15] as well as of other groups [10-12] – see the lone data point at 10^{-3} bar in Fig. 3.10.

Below 10^{-4} bar, however, this measurement yields a smaller slope. The first decomposition process is determined between 10^{-12} bar and 10^{-14} bar in both measurements. The measurements in this paper were carried out on powder samples with small particle sizes and, hence, small oxygen diffusion lengths, whereas Mueller et al. [15] used densely sintered ceramic bulk samples. As no detailed kinetic aspects of the measurement have been reported in [15], it is hard to determine whether the dense sample reached equilibrium for each data point.

Owing to the highly complex nature of such a titration method, several fallacies may easily occur: The most important aspect is taking the kinetics of the system into account. This requires a sufficiently fast sample equilibration at each set pO_2 value, and also a reliable determination of the actual pO_2, preferably *continuously* and *in situ*. Elsewise, the measured pump current may not necessarily reflect the *reversible* $\Delta\delta$ values of the sample as, e.g., pumping oxygen out of a sample chamber with a very small dead volume may quickly lead to pO_2 non-equilibria: If oxygen removal takes place much faster than the oxygen equilibration of a massive sample, then a pO_2 value may temporarily occur near the sample surface that is more reducing than assumed (and measured after the pumping procedure). Hence, this might lead to a decomposition front moving into the material already at apparently higher pO_2 values and, thus, the $\Delta\delta$ values determined from the pump current would not stem from *reversible* oxygen exchange but possibly rather from the already beginning sample decomposition. Titration measurements performed on powder samples with a sufficient amount of surrounding gas phase can clearly avoid this pitfall.

3.3.5 Decomposition Products of BSCF

In order to examine the decomposition products present in BSCF after being exposed to very low oxygen partial pressures, X-ray diffractometric analyses have been conducted after annealing powder for 10 hours at the low partial pressure of 10^{-15} bar, as well as a powder sample that underwent the symmetrical titration back up to ambient air pressure, as indicated by the blue arrows in Fig. 3.11.

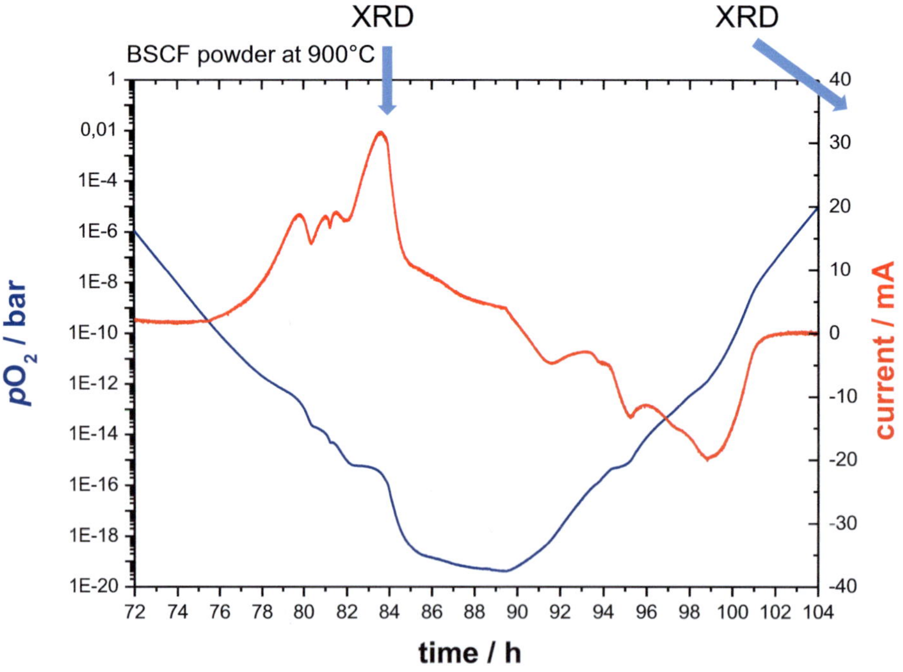

Fig. 3.11: Symmetrical titration curve of BSCF powder at 900 °C. Blue arrows indicate that XRD measurements were conducted on powder annealed at 10^{-15} bar as well as on a powder batch after the symmetrical titration was finished.

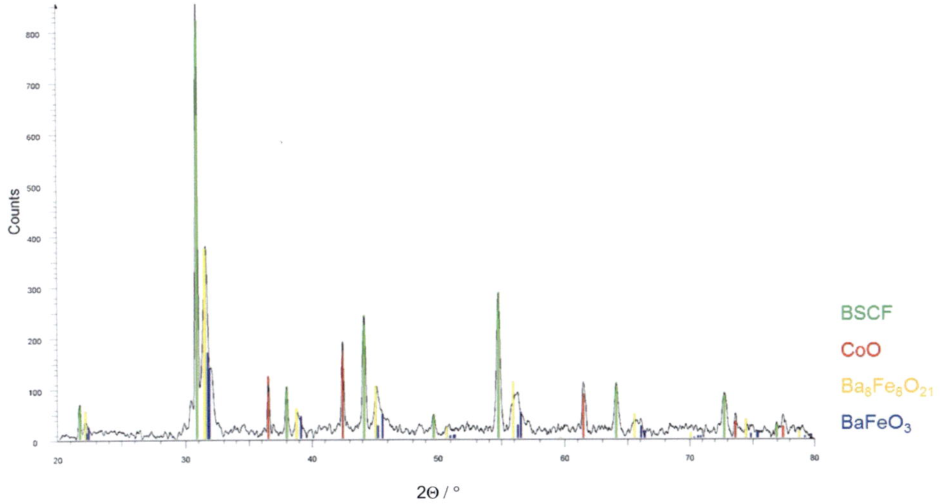

Fig. 3.12: XRD measurement of powder after coulometric titration at 900 °C down to 10^{-15} bar pO_2 and annealing at that point for 10 hours.

The first XRD measurement of powder after coulometric titration down to 10^{-15} bar pO_2 and annealing at that point for 10 hours is shown in Fig. 3.12. Prominently visible is the presence of cobalt monoxide CoO indicated by the red peak positions. The remaining peaks present in this measurement can be attributed to a remaining cubic BSCF lattice (green) as well as other cubic lattices with slightly different lattice parameters that have been indexed by the software as barium ferrites, but should be barium strontium ferrites not present in the XRD database. The initial BSCF matrix is destroyed by emanating CoO, leaving different cubic lattices of various compositions.

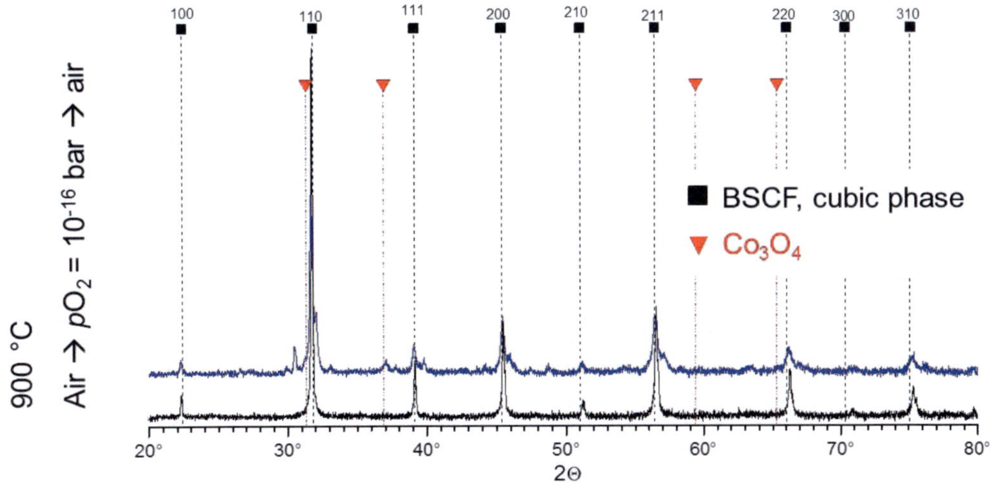

Fig. 3.13: XRD measurement of BSCF powder after completed symmetrical coulometric titration at 900 °C from air to $pO_2 = 10^{-16}$ bar and back (blue) in comparison to untreated powder (black).

After a symmetrical coulometric titration run down to 10^{-16} bar pO_2 at 900 °C, the different cubic decomposition products are still present (cf. Fig. 3.13), the only significant change being the additional presence of Co_3O_4, which is an expected change, as the phase diagram of cobalt oxide suggests that composition to form at 0.21 bar at 900 °C.

By annealing fresh BSCF powders in the oxygen pump setup for 10 hours at different temperatures of 700, 800 and 900 °C, a stability matrix for the BSCF system could be obtained as shown in Table 3.1:

	700 °C	**800 °C**	**900 °C**
10^{-16} bar	not stable	not stable	not stable
10^{-15} bar	not stable	not stable	not stable
10^{-14} bar	stable	not stable	not stable
10^{-13} bar	stable	stable	not stable
10^{-12} bar	stable	stable	stable

Tab. 3.1: Stability matrix for BSCF powder when annealing for 10 hours at the given temperature (top row), depending on oxygen partial pressure (left column).

3.3.6 Stability Limits of other Perovskites

Similar coulometric titration measurements have been conducted on powders of LSCF, LSC and PSCF, partly shown in our paper [1]. Fig. 3.14 shows the coulometric titration curve of an LSCF powder sample at $T = 900$ °C ($dU_{pump}/dt = 1$ mV/min) for the pO_2 range below 10^{-6} bar. By monitoring the pump current low-pO_2 decomposition processes are observed (in reproducible experiments) where oxygen is set free from the sample (around $3 \cdot 10^{-13}$ and $2 \cdot 10^{-14}$ bar). One can clearly distinguish two regimes: At pO_2 values above ca. $3 \cdot 10^{-13}$ bar LSCF reversibly excorporates gaseous oxygen. In this plot the deviation in actual pO_2 within the sample chamber (resulting from this oxygen excorporation) from the set value (dashed blue line) becomes clearly visible below ~10^{-10} bar. At pO_2 values of around $3 \cdot 10^{-13}$ bar the maximum of the first peak in I_{pump} is reached (first shaded red arrow); this heralds the low-pO_2 stability limit; at even lower pressures the LSCF is no longer chemically stable. The "buffering" effect of this chemical process is also visible in the interplay between current peak and transient pO_2 deviation: when the former is completed, the latter strives at regaining the preset pO_2 value again.

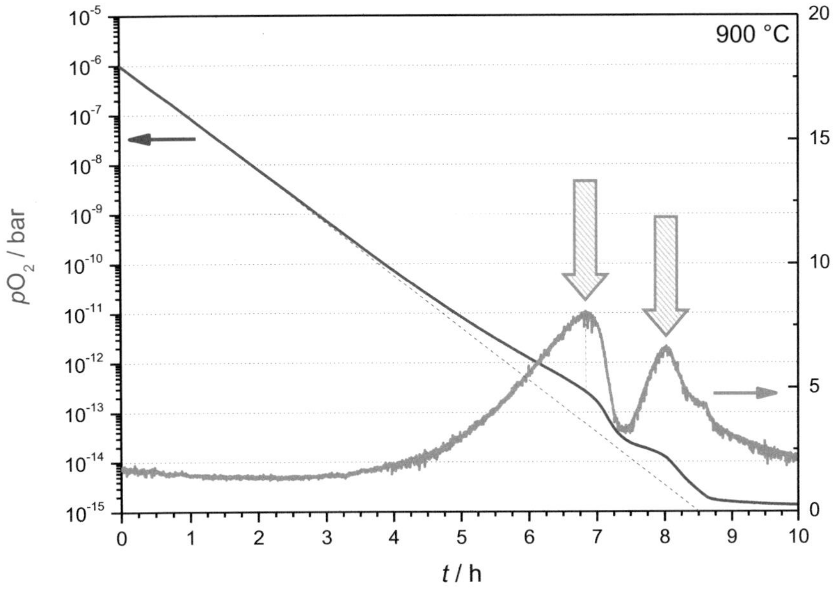

Fig. 3.14: Coulometric titration (dU_{pump}/dt = 1 mV/min) of an LSCF powder sample at $T = 900$ °C from 10^{-6} bar down to very low oxygen partial pressures.

This result compares well to the data shown on BSCF and also on LSC (Fig. 3.15) as well as data measured on PSCF (Fig. 3.16). For the LSC in Fig. 3.15, one can also clearly distinguish two regimes: At pO_2 values above ca. 10^{-13} bar LSC reversibly excorporates gaseous oxygen.

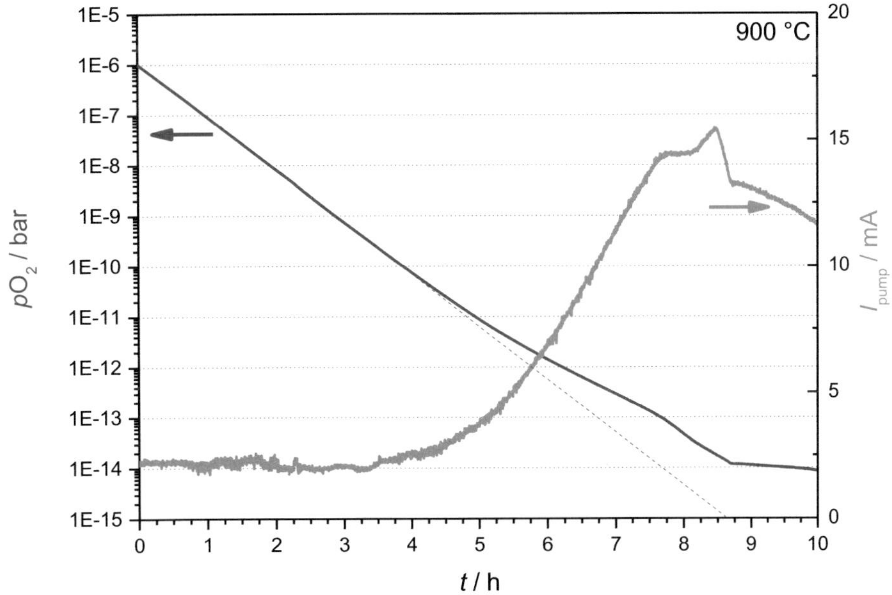

Fig. 3.15: Coulometric titration (dU_{pump}/dt = 1 mV/min) of an LSC powder sample at $T = 900$ °C from 10^{-6} bar down to very low oxygen partial pressures.

In this plot the deviation in actual pO_2 within the sample chamber (resulting from this oxygen excorporation) from the set value (dashed blue line) becomes clearly visible again below ~10^{-10} bar. At pO_2 values of around 10^{-13} bar a first maximum of I_{pump} is reached; this heralds the low-pO_2 stability limits. Compared to the case of LSCF (Fig. 3.14) the current peak appears much more pronounced, indicating a faster decomposition.

For the PSCF powder titration shown in Fig. 3.16, the same two regimes are visible: At pO_2 values above ca. 10^{-10} bar PSCF reversibly excorporates gaseous oxygen. This time, the deviation in actual pO_2 within the sample chamber (resulting from this oxygen excorporation) from the set value (dashed blue line) starts slightly below 10^{-9} bar. At pO_2 values of around 10^{-11} bar a maximum in I_{pump} is reached indicating the low-pO_2 stability limit. Compared to the case of LSCF (Fig. 3.14) the current peak appears much shallower, indicating a slower decomposition.

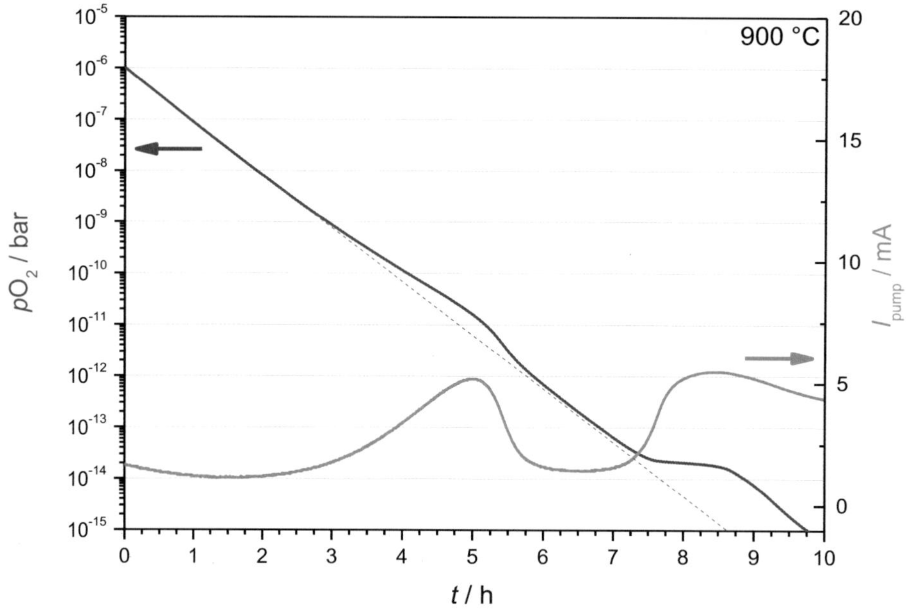

Fig. 3.16: Coulometric titration (dU_{pump}/dt = 1 mV/min) of a PSCF powder sample at T = 900 °C from 10^{-6} bar down to very low oxygen partial pressures.

Table 3.2 summarizes the results for all four MIEC compositions. While the stability limits determined by coulometric titration are fairly similar in all four cases, there are notable differences in the decomposition kinetics (as can be seen in the different time dependencies of the pump current), that are correlated with the B-site Co/Fe ratio. Higher iron content appears to slow down the decomposition kinetics.

Stability at 900 °C	LSCF	PSCF	BSCF	LSC
Stable down to pO_2 =	$3 \cdot 10^{-13}$ bar	$3 \cdot 10^{-11}$ bar	$5 \cdot 10^{-13}$ bar	$\sim 10^{-13}$ bar
Decomposition kinetics in powders	slow	slow	fast	very fast

Tab. 3.2: Summary of results on low-pO_2 stability (at 900 °C) of the different MIEC compositions investigated by coulometric titration experiments. Decomposition kinetics are determined by an analysis of the first current peak in the coulometric titration curve.

3.4 Conclusions

In this chapter, a refined closed tubular zirconia "oxygen pump" setup with Pt electrodes is shown that facilitates precise control of the oxygen partial pressure in the enclosed volume within a pO_2 range between 10^{-18}...1 bar at temperatures above 700 °C. Oxygen uptake or release of an oxide powder sample can be determined with high precision as a function of pO_2, temperature and time, yielding information on stoichiometry changes and stability limits of an MIEC oxide, by:

- measuring the electric pump current,
- correction for intrinsic setup properties through an offset or background subtraction,
- continuous in-situ determination of the oxygen partial pressure at the sample position with high accuracy (validated on an electrically contacted $SrTiO_3$ single-crystal sample),
- adjusting the pO_2 change rate to resolve kinetic behavior,
- providing convenient experimental simplicity (re-usability, easy sample exchange, sealing at room-temperature).

This method is validated with the help of the "model" oxide CuO where pO_2-dependent transition conditions are known from phase diagrams: The reduction of CuO to Cu_2O could be observed via a clearly visible buffering effect at $pO_2 \approx 5 \cdot 10^{-3}$ bar (at 850 °C, in good agreement with [21]). The amount of oxygen transported during the reduction reaction could be determined with only a 6 % deviation from the theoretical amount.

In the next step, this coulometric titration method was applied to more complex MIEC materials, such as $Ba_{0.5}Sr_{0.5}Co_{0.8}Fe_{0.2}O_{3-\delta}$ (BSCF), thus yielding information on oxygen stoichiometry changes and irreversible demixing:

- Several processes become visible by this method: Reversible oxygen exchange with the ambient atmosphere as well as irreversible demixing at very low pO_2.

- Oxygen non-stoichiometry changes ($\Delta\delta$) can be calculated for a wide pO_2 range, a significant drop between 10^{-11} bar and 10^{-13} bar indicates the end of the stability range for BSCF (at 900 °C).
- These stability limits have been verified by XRD analyses.

By employing the continuous coulometric titration technique the low-pO_2 stability limits of other selected application-relevant MIEC powder samples could also conveniently be determined:

- All investigated compositions ($La_{0.58}Sr_{0.4}Co_{0.2}Fe_{0.8}O_{3-\delta}$ (LSCF), $Pr_{0.58}Sr_{0.4}Co_{0.2}Fe_{0.8}O_{3-\delta}$ (PSCF), $Ba_{0.5}Sr_{0.5}Co_{0.8}Fe_{0.2}O_{3-\delta}$ (BSCF), and $La_{0.6}Sr_{0.4}CoO_{3-\delta}$ (LSC)) proved to be stable down to (at least) $pO_2 \approx 10^{-10}$ bar at 900 °C.
- This pO_2 stability range is sufficient for most applications (SOFC cathode, oxygen-transport membrane).
- With increasing perovskite B-site iron-to-cobalt ratio, decomposition kinetics appear to slow down.

References

[1] C. Niedrig, S. F. Wagner, W. Menesklou and E. Ivers-Tiffée, "Characterization of oxygen dependent stability of selected mixed-conducting perovskite oxides", *Solid State Ionics* **273**, pp. 41-45 (2015).

[2] C. Niedrig, W. Menesklou, S. F. Wagner and E. Ivers-Tiffée, "High-Temperature pO_2 Stability of Metal Oxides Determined by Amperometric Oxygen Titration", *Journal of The Electrochemical Society* **160**, pp. F135-F140 (2013).

[3] C. Niedrig, P. Braun, W. Menesklou, S. F. Wagner and E. Ivers-Tiffée, "Investigation of MIEC Materials using an Amperometric Oxygen Titration Method", *ECS Transactions* **45**, pp. 75-85 (2012).

[4] S. F. Wagner, S. Taufall, C. Niedrig, H. Götz, W. Menesklou, S. Baumann and E. Ivers-Tiffée, "pO_2 stability of $Ba_{0.5}Sr_{0.5}Co_{0.8}Fe_{0.2}O_{3-\delta}$", *Materials Research Society Symposium Proceedings* **1309**, pp. 57-64 (2011).

[5] S. Švarcová, K. Wiik, J. Tolchard, H. J. M. Bouwmeester and T. Grande, "Structural instability of cubic perovskite $Ba_xSr_{1-x}Co_{1-y}Fe_yO_{3-\delta}$", *Solid State Ionics* **178**, pp. 1787-1791 (2008).

[6] D. N. Müller, R. A. De Souza, T. E. Weirich, D. Roehrens, J. Mayer and M. Martin, "A kinetic study of the decomposition of the cubic perovskite-type oxide $Ba_xSr_{1-x}Co_{0.8}Fe_{0.2}O_{3-\delta}$ (BSCF) (x=0.1 and 0.5)", *Physical Chemistry Chemical Physics* **12**, pp. 10320-10328 (2010).

[7] K. Efimov, Q. Xu and A. Feldhoff, "Transmission Electron Microscopy Study of $Ba_{0.5}Sr_{0.5}Co_{0.8}Fe_{0.2}O_{3-\delta}$ Perovskite Decomposition at Intermediate Temperatures", *Chemistry of Materials* **22**, pp. 5866-5875 (2010).

[8] C. Niedrig, S. Taufall, M. Burriel, W. Menesklou, S. F. Wagner, S. Baumann and E. Ivers-Tiffée, "Thermal stability of the cubic phase in $Ba_{0.5}Sr_{0.5}Co_{0.8}Fe_{0.2}O_{3-\delta}$ (BSCF)", *Solid State Ionics* **197**, pp. 25-31 (2011).

[9] P. Müller, H. Störmer, L. Dieterle, C. Niedrig, E. Ivers-Tiffée and D. Gerthsen, "Decomposition pathway of cubic $Ba_{0.5}Sr_{0.5}Co_{0.8}Fe_{0.2}O_{3-\delta}$ between 700 °C and 1000 °C analyzed by electron microscopic techniques", *Solid State Ionics* **206**, pp. 57-66 (2012).

[10] S. McIntosh, J. F. Vente, W. G. Haije, D. H. A. Blank and H. J. M. Bouwmeester, "Oxygen stoichiometry and chemical expansion of $Ba_{0.5}Sr_{0.5}Co_{0.8}Fe_{0.2}O_{3-\delta}$ measured by in situ neutron diffraction", *Chemistry of Materials* **18**, pp. 2187-2193 (2006).

[11] E. Bucher, A. Egger, P. Ried, W. Sitte and P. Holtappels, "Oxygen nonstoichiometry and exchange kinetics of $Ba_{0.5}Sr_{0.5}Co_{0.8}Fe_{0.2}O_{3-\delta}$", *Solid State Ionics* **179**, pp. 1032-1035 (2008).

[12] R. Kriegel, R. Kircheisen and J. Töpfer, "Oxygen stoichiometry and expansion behavior of $Ba_{0.5}Sr_{0.5}Co_{0.8}Fe_{0.2}O_{3-\delta}$", *Solid State Ionics* **181**, pp. 64-70 (2010).

[13] S. McIntosh, J. F. Vente, W. G. Haije, D. H. A. Blank and H. J. M. Bouwmeester, "Structure and oxygen stoichiometry of $SrCo_{0.8}Fe_{0.2}O_{3-\delta}$ and $Ba_{0.5}Sr_{0.5}Co_{0.8}Fe_{0.2}O_{3-\delta}$", *Solid State Ionics* **177**, pp. 1737-1742 (2006).

[14] J. Ovenstone, J. I. Jung, J. S. White, D. D. Edwards and S. T. Misture, "Phase stability of BSCF in low oxygen partial pressures", *Journal of Solid State Chemistry* **181**, pp. 576-586 (2008).

[15] D. N. Mueller, R. A. De Souza, H.-I. Yoo and M. Martin, "Phase Stability and Oxygen Nonstoichiometry of Highly Oxygen-Deficient Perovskite-Type Oxides: A Case Study of $(Ba,Sr)(Co,Fe)O_{3-\delta}$", *Chemistry of Materials* **24**, pp. 269-274 (2012).

[16] K. Beetz, *Die geschlossene Festelektrolyt-Sauerstoffpumpe*, Düsseldorf: VDI Verlag GmbH (1993).

[17] J. Mizusaki, H. Tagawa, K. Naraya and T. Sasamoto, "Nonstoichiometry and thermochemical stability of the pervoskite-type $La_{1-x}Sr_xMnO_{3-\delta}$", *Solid State Ionics* **49**, pp. 111-118 (1991).

[18] B. Zachau-Christiansen, T. Jacobsen and S. Skaarup, "Eletrochemical determination of oxygen stoichiometry and entropy in oxides", *Solid State Ionics* **86-88**, pp. 725-731 (1996).

[19] S.-H. Kang and H.-I. Yoo, "Nonstoichiometry (δ) of $(Mg_x,Fe_{1-x})_{3-\delta}O_4$, ferrite spinel", *Solid State Ionics* **86-88**, pp. 751-755 (1996).

[20] B. T. Dalslet, M. Søgaard, H. J. M. Bouwmeester and P. V. Hendriksen, "Defect chemistry and oxygen transport of $(La_{0.6}Sr_{0.4-x}M_x)_{0.99}Co_{0.2}Fe_{0.8}O_{3-\delta}$, M=Ca (x=0.05, 0.1), Ba (x=0.1, 0.2), Sr", *Solid State Ionics* **180**, pp. 1173-1182 (2009).

[21] J. P. Neumann, T. Zhong and Y. A. Chang, "The Cu-O (Copper-Oxygen) System", *Journal of Phase Equilibria* **5**, pp. 136-140 (1984).

[22] J. H. Park and R. N. Blumenthal, "Electronic transport in 8 mole percent Y_2O_3-ZrO_2", *Journal of The Electrochemical Society* **136**, pp. 2867-2876 (1989).

[23] J. Werner and R. Schmid-Fetzer, "Performance of ZrO_2 Electrolyte Cells for Coulometric Titration at High Voltage", *Thermochimica Acta* **129**, pp. 127-141 (1988).

[24] P. R. Shah, T. Kim, G. Zhou, P. Fornasiero and R. J. Gorte, "Evidence for Entropy Effects in the Reduction of Ceria-Zirconia Solutions", *Chemistry of Materials* **18**, pp. 5363-5369 (2006).

[25] S. Ganschow, D. Schulz, D. Klimm, R. Bertram and R. Uecker, "Application of predominance diagrams in melt growth of oxides", *Crystal Research and Technology* **45**, pp. 1219-1224 (2010).

[26] J. Assal, B. Hallstedt and L. J. Gauckler, "Experimental Phase Diagram Study and Thermodynamic Optimization of the Ag-Bi-O System", *Journal of the American Ceramic Society* **82**, pp. 711-715 (1999).

[27] O. Kubaschewski, R. Schmid-Fetzer, L. Rokhlin, L. Cornish and O. Fabrichnaya, "Aluminium - Iron - Oxygen", in G. Effenberg and S. Ilyenko (Eds.), *Landolt-Börnstein - Group IV Physical Chemistry*, Vol. 11D1: Iron Systems, Part 1, Berlin, Heidelberg: Springer-Verlag, pp. 1-48 (2008).

[28] E. Passerone, R. Ricci and R. Sangiorgi, "Influence of oxygen contamination on the surface tension of liquid tin", *Journal of Materials Science* **25**, pp. 4266-4272 (1990).

[29] M. Brust, *In situ Charakterisierung von Vanadiumoxid/Titanoxid-Katalysatoren bei der partiellen Oxidation von o-Xylol zu Phtalsäureanhydrid*, Dissertation, Fakultät für Chemieingenieurwesen, Universität Karlsruhe (TH) (1999).

[30] E. Schleitzer-Steinkopf, "Verbindungen des Mangans mit Sauerstoff und Metallen der 3. bis 6. Gruppen des Periodensystems: Mangan-Stickstoff-Verbindungen", in E. Schleitzer-Rust (Ed.), *Gmelin-Handbuch der anorganischen Chemie*, Berlin, Heidelberg, New York: Springer Verlag, pp. 1-307 (1975).

[31] R. Moos and K. H. Härdtl, "Defect chemistry of donor-doped and undoped strontium titanate ceramics between 1000° and 1400 °C", *Journal of the American Ceramic Society* **80**, pp. 2549-2562 (1997).

Chapter 4

Conductivity of BSCF over Temperature

4.1 Introduction

For mixed ionic electronic conducting oxygen permeation membranes, higher fluxes are obtained when either the electronic or ionic conductivity increases, or both increase simultaneously – as permeation rate depends on the product of mobility and concentration of both ionic and electronic charge carriers. Furthermore, for a sufficiently thick membrane, oxygen permeation will be controlled by bulk diffusion, which is why reducing the membrane thickness results in an increased oxygen flux. To this end, a common proposed solution is to apply a supported membrane: a dense thin film on a porous substrate [3-6]. However, this applies only until reaching thicknesses of a certain value (the "characteristic" membrane thickness typical of the material employed [7], given by the ratio of oxygen diffusion constant and surface exchange coefficient), when the diffusion processes become sufficiently fast, resulting in a change in the kinetic behavior. The surface reactions will finally become the rate determining step for such very thin membranes [3]. No enhancement of the oxygen flux can be achieved by thinning the membrane any further if the surface oxygen exchange cannot be increased simultaneously.

To investigate the BSCF system regarding its electrical conductivity, this chapter discusses conductivity measurements on sintered bulk samples and thin films (epitaxial thin films obtained by Pulsed Laser Deposition (PLD) and polycrystalline thin films obtained by Metal-Organic Deposition (MOD) through spin coating, both on NGO substrates) in the temperature range from room temperature up to 900 °C.

Parts of the results shown in this chapter have been published in [1;2].

Electrical conductivity properties of BSCF measured on polycrystalline dense bulk samples can be compared to available data measured in air by Zhou et al. and Wei et al. [8;9], and in air and oxygen atmosphere by Chen et al. [10].

Thin films become important on one hand due to the increased oxygen permeation through very thin membranes mentioned earlier. On the other hand, the transport properties of polycrystalline samples are strongly affected by the sample morphology, such as particle size, grain boundary shape, connectivity, or porosity, making the intrinsic properties of BSCF difficult to extract. Investigating epitaxial samples avoids these detrimental influences, while MOD thin films are interesting in regard to the much simpler production process. Measurements regarding the oxygen surface exchange coefficient k^{δ} are discussed in Chapters 6 and 7 for sintered bulk samples and thin films, respectively.

4.2 Experimental

4.2.1 Contacting of the Samples

For the conductivity measurements two parallel gold contacts were painted at the surface of the epitaxial PLD-derived thin-film samples (preparation cf. Chapter 2) and fired for one hour at a temperature of 950 °C in air. The MOD films were contacted with two parallel gold contacts by magnetron sputtering and subsequent spot-welding of Au-wires. Owing to the fairly high electrical resistance of the thin films, their planar resistance was measured in two-point-technique in the oxygen pump setup (cf. Chapters 2, 3).

Bulk samples were measured in four-point-technique in the same setup; contacts were applied by means of Pt wires and paste for sample FZJ-BSCF-02, and by Pt magnetron sputtering, Pt wires and paste for all other shown samples, with subsequent firing for one hour at 1050 °C in air.

4.3 Results and Discussion

4.3.1 Sintered Bulk Samples

The electrical properties of BSCF bulk samples have been measured in pure oxygen atmosphere and in air in the temperature range from 40 °C to 900 °C while cooling down from the highest temperature at a rate of 1 K/min.

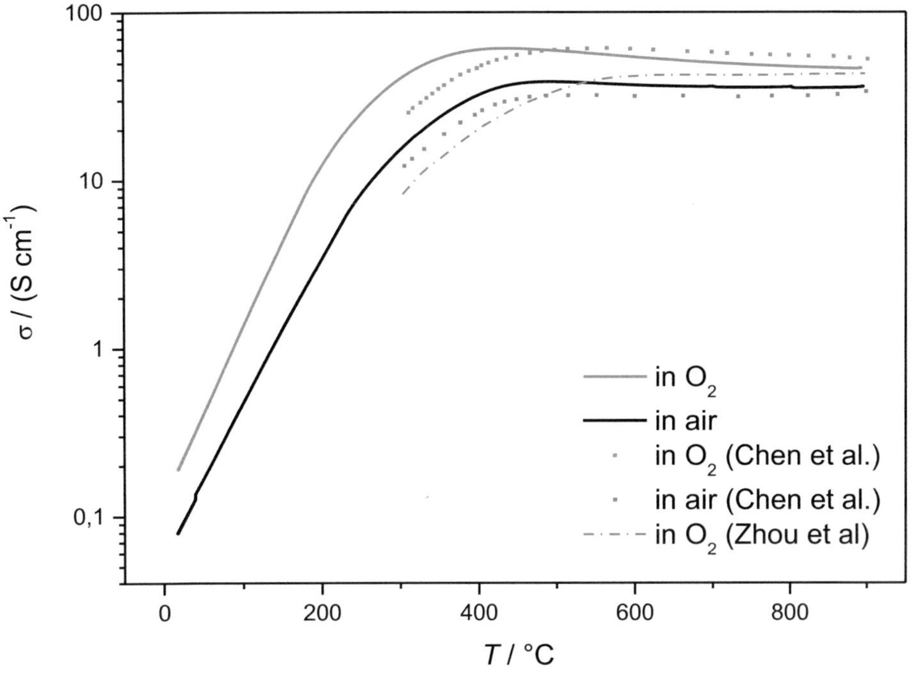

Fig. 4.1: Conductivity of BSCF sample FZJ-BSCF-02 during cooling in oxygen (red curve) and air (black curve) in comparison to data published by Chen et al. [10] and Zhou et al. [8] as published in [1].

Fig. 4.1 shows the conductivity values obtained from bulk sample measurements both from this work and from literature data. The conductivity of BSCF increases with temperature, presenting a semiconductor-type behavior corresponding to a small polaron hopping mechanism, associated with the behavior of mixed (2+/3+) valence states of cobalt and iron cations. It can be observed that the behavior of the BSCF bulk sample nicely matches the data published by other authors for BSCF polycrystalline samples [8;10]. The conductivity increases with temperature up to approximately 400 °C, then reaching a plateau. At high temperatures the creation of additional oxygen vacancies, accompanied by a reduction of Fe^{4+} to Fe^{3+} or Co^{3+} to Co^{2+}, causes a decrease in the charge carrier concentration and a covalent interaction because of a perturbation of the O–(Fe,Co)–O periodic potential, which may account for such a transition [8].

The behavior of the bulk sample in pure oxygen and in air have been plotted in an Arrhenius-type graph in Fig. 4.2, showing p-type conductivity ($\sigma_{\text{oxygen}}>\sigma_{\text{air}}$). The corresponding values of the activation energy (E_a) have been calculated from the representation of σT vs. $1/T$ by the adiabatic small polaron hopping model:

$$\sigma = \frac{A_o}{T}\exp\left(-\frac{E_a}{RT}\right) \tag{1}$$

where σ is the electrical conductivity, A_o is a material constant containing the carrier concentration term, T is the absolute temperature, E_a is the activation energy for small polaron hopping, and R is the universal gas constant.

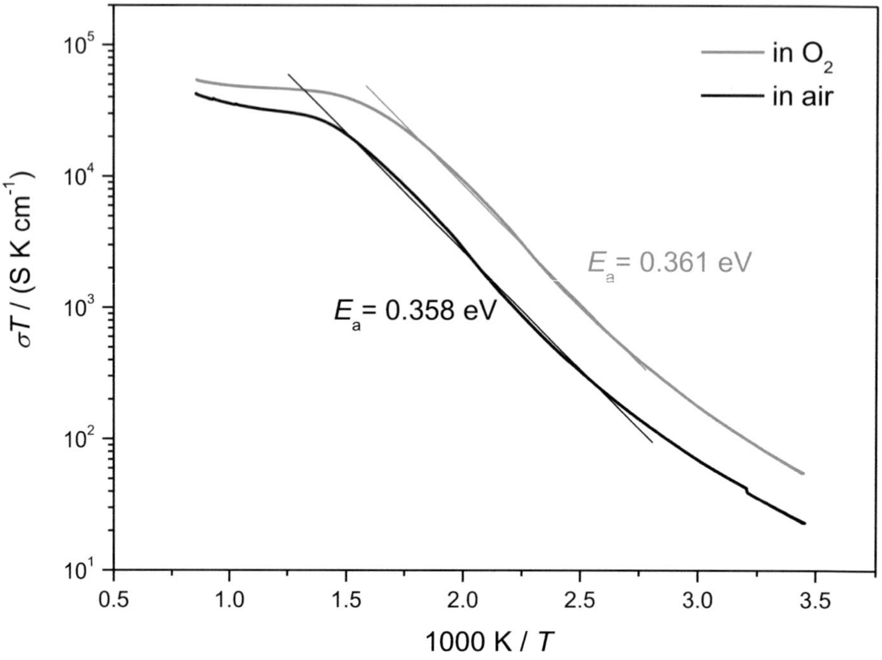

Fig. 4.2: Activation energy of the electrical conductivity of the BSCF sample FZJ-BSCF-02 in oxygen (red curve) and air (black curve) as published in [1].

In both atmospheres (pure oxygen and air), activation energies are quite similar, calculated as 0.361 eV and 0.358 eV respectively.

Generally, the conductivity is higher in pure oxygen atmospheres than it is in air, owing to the fact that in pure oxygen, more oxygen ions are introduced into the BSCF lattice, reducing the inherent oxygen nonstoichiometry and by requiring two electrons each creating more holes as charge carriers in the p-type conductor. As these measurements are done while slowly cooling down from the highest temperature, the different stoichiometries are frozen at low temperatures below 300 °C to 400 °C where oxygen exchange no longer takes place. The conductivity curves stay parallel to each other even at these temperatures, resulting in different conductivity values at room temperature. Heating curves of identical bulks with similar thermal history would yield identical conductivity values up to these temperatures and then the measurement in pure oxygen would start deviating to the depicted higher conductivity values starting in this temperature range with the onset of oxygen exchange in the sample.

4.3.2 Epitaxial Thin Films

Under real operating conditions, during the heating and cooling processes a membrane experiences, the membrane material will not be able to expand freely, but will be restricted by a thicker porous substrate, in the case of a supported membrane, and by the thermal expansion coefficient of the module construction materials (e.g. steel). For BSCF epitaxial thin films on single crystal perovskite substrates, similar restrictions apply. Due to the differences in cell parameters of BSCF and substrate and resulting stress, the oxygen stoichiometry, vacancy concentration and/or the valance state of the B-site metal ions of the BSCF perovskite could be different to those of a polycrystalline material (subjected to less expansion limitations), and could affect the electronic conductivity of the film.

For comparison, the electrical properties of a 340 nm-thick BSCF epitaxial film on NGO substrates have also been measured under the same atmospheric conditions in the same temperature range. It should be noted that the conductivity of the substrate ($\sigma_{NGO,\,700\,°C} \approx 10^{-6}$ S cm^{-1}) is several orders of magnitude smaller than that of the layers ($\sigma_{BSCF,\,700\,°C} \approx 10^{2}$ S cm^{-1}). So, even though the substrate (d = 0.5 mm) is more than a thousand times thicker than the BSCF thin films, its conductance is still negligible, ensuring that the layer conductivity is measured without influence from the substrate. In Figure 4.3 the total conductivity of the thin film sample measured in air has been plotted vs. temperature in comparison to the bulk sample conductivities (left), while data in both atmospheres is shown in an Arrhenius graph on the right.

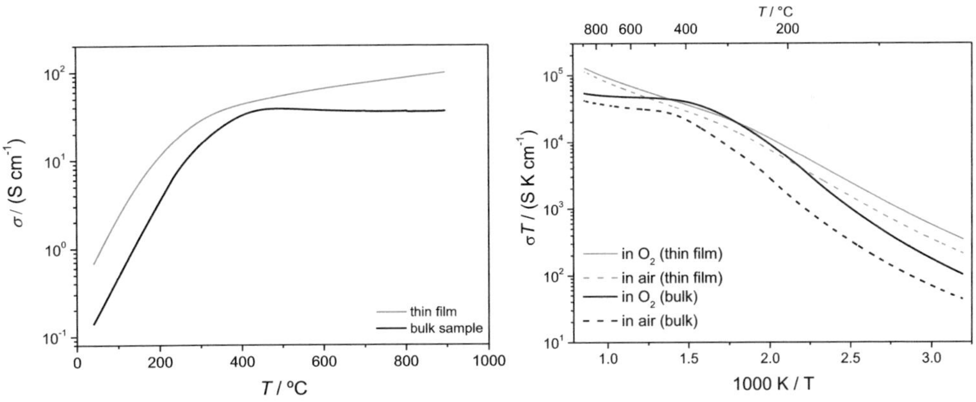

Fig. 4.3: Left: comparison of the electronic conductivity in air of the BSCF bulk sample FZJ-BSCF-02 (black curve) and epitaxial thin film PLD-BSCF-17n2 (red curve) and right: the Arrhenius graph for both samples in air (dashed curves) and oxygen (solid curves), where bulk conductivity is shown in black, thin film conductivity in red [1].

Although the conductivity of the thin film also increases with temperature and presents a typical semiconductor-type behavior it performs slightly differently. Firstly, the overall conductivity of the epitaxial thin film in air is between two and five times higher in the whole temperature range measured. Furthermore, while for the thin film a similar behavior is found at low temperatures below approx. 400 °C, at higher temperatures the conductivity does not reach a maximum, but continues to increase slightly in the whole temperature range. This could be due to differences in the oxygen vacancies content and defects equilibration stoichiometry. Considering that the conductivity curve deviation at higher temperatures is only related to the loss of oxygen by surface exchange, the oxidation enthalpy seems to be considerably reduced in thin films compared to that in bulks, which could be related to the stress induced by the substrate and/or to the crystallographic orientation of the surface planes. The conductivity of the thin film exhibits a similar behavior in oxygen atmosphere and in air, increasing from the value of 0.68 S cm^{-1} for oxygen (0.14 S cm^{-1} for air) at 40 °C to 110.5 S cm^{-1} for oxygen (98.5 S cm^{-1} for air) at 900 °C, and consisting of two different temperature ranges with different activation energies. In both ranges the conductivity shows an Arrhenius-type dependence with temperature with a change in slope at a transition temperature of about 300 °C. In the case of bulk samples the activation energies at low temperatures are higher than in the thin films and the transition temperature between the two ranges occurs at a higher temperature of about 400 °C. This difference may be attributed to the absence of grain boundaries and the stress introduced into the thin film by the substrate. The activation energy values of the bulk calculated for both gas atmospheres are summarized in Table 4.1 and compared to the thin film's activation energy values in both regions.

Gas atmosphere	**E_a (eV)**		
	Bulks	**Thin films**	
	40-400 °C	**40-300 °C**	**300-900 °C**
Pure oxygen	0.361	0.243	0.168
Air	0.358	0.254	0.182

Tab. 4.1: Activation energies of the conductivity (σT) for the BSCF bulks and thin films [1].

4.3.3 Polycrystalline MOD Thin Films

Due to their much simpler – and on a large scale cheaper – production process, MOD derived polycrystalline thin films are also a matter of interest. An 80 nm thick BSCF film on NGO prepared by K. Asano (cf. Chapter 2) has been measured in air with the same cooling rate as the previous samples. Fig. 4.4 shows the conductivity vs. temperature of that film in comparison to both the discussed bulk sample data and the results for the epitaxial thin film produced by PLD.

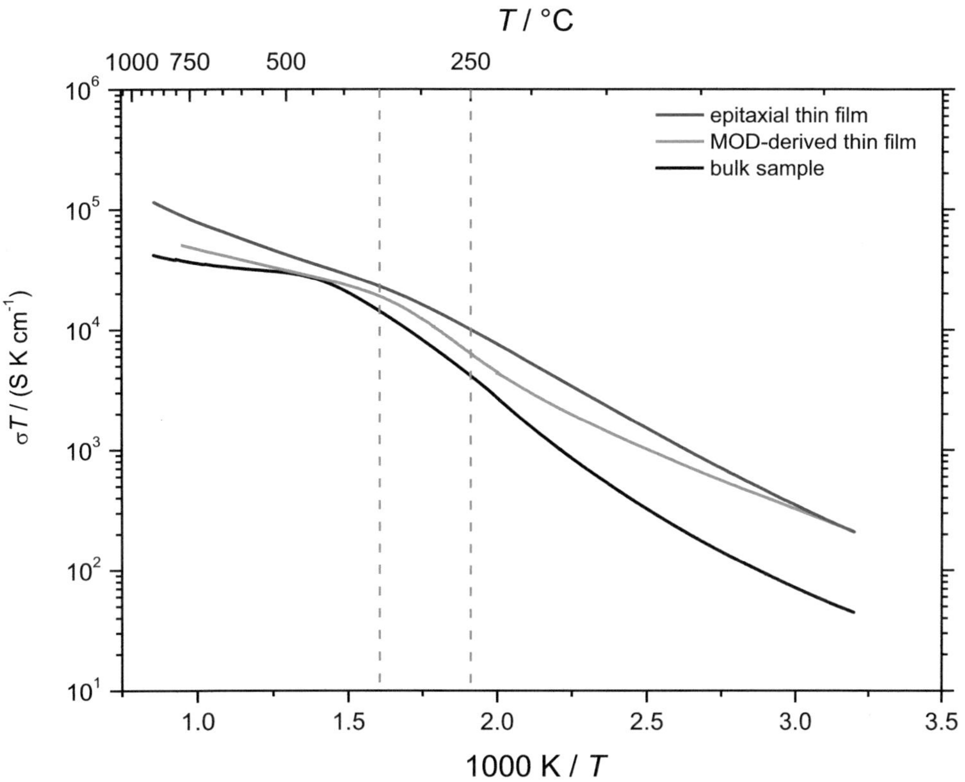

Fig. 4.4: Comparison of the electrical conductivity of BSCF bulk samples (FZJ-BSCF-02, black curve), epitaxial thin films (PLD-BSCF-17n2, blue curve) and MOD-derived thin films (MOD-BSCF-066, red curve) [2].

In this measurement, three different regimes become visible. Starting at high temperatures above 400 °C, the MOD film's conductivity values range between those for the bulk – which is lowest due to the unrestrained ability to emit oxygen into the surrounding atmosphere – and epitaxial film – with the highest value due to the absence of grain boundaries and the stress introduced by the substrate. Between approx. 350 °C and 250 °C, the activation energy of the MOD film is quite similar to that of the bulk sample, while that of the PLD film is lower, suggesting that in this temperature range, thermal and chemical behavior is comparable for the first two mentioned samples due to the presence of grain boundaries, with which slight shifts in lattice constants are easier to handle for the system. Finally below 250 °C, where oxygen stoichiometry is frozen due to the low temperatures, the MOD film's conductivity decreases with the flattest slope to finally reach the same conductivity value as the epitaxial film at room temperature, suggesting that this higher value – compared to the unrestrained bulk sample – is proof for the fact that the substrate induced stress on the material retains the BSCF thin films from emitting as much oxygen towards a higher vacancy concentration as the bulk sample is able to do. This is due to the smaller pseudo-cubic lattice constant of the NGO substrate.

4.3.4 Doped BSCF Bulk Samples

In order to stabilize the cubic phase in BSCF at the intended operating temperatures of approx. 800 °C, several monovalent substituents on the B-site have been proposed as mentioned in Chapters 1, 2 and given an in-depth examination in Chapter 8. To examine their individual performance, measuring the electrical conductivity is one of the methods of choice. Fig. 4.5 shows conductivity values during cooling down measurements as shown in the earlier parts of this chapter for several pure BSCF bulk samples in comparison to samples that have been doped with 3 at% of Zr, Y, and Sc on the B-site, respectively.

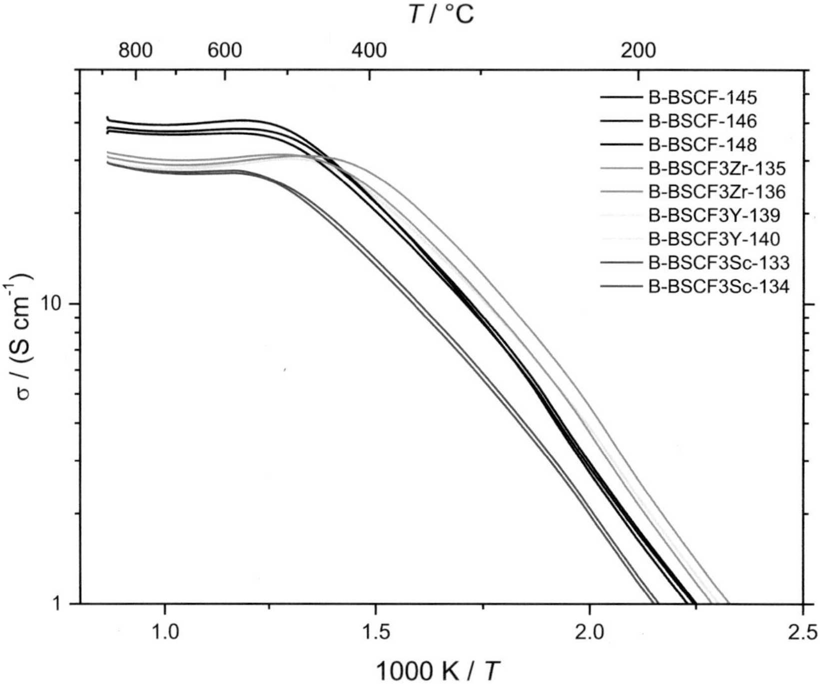

Fig. 4.5: Electrical conductivity of doped BSCF samples during cooling in air for BSCF3Zr (red curves), BSCF3Y (green curves) and BSCF3Sc (blue curves) in comparison to BSCF (black curves).

As there are three measurements on different samples of undoped BSCF and in the case of each dopant two measurements of different bulk samples, a slight deviation for the measurements of each composition can be seen. This is on one hand due to the contacting of the samples with sputtered Pt contacts, Pt paste and Pt wires (cf. Chapter 2) that introduces some inherent error into the conductivity calculation as well as different achieved densities of the samples. While the first can be handled by statistical mean value calculation, the second can be corrected for if the porosity of the samples is known. This second point is also the reason that, from the graphs in Fig. 4.5, no conclusive statement can be made on the influence of different dopants in comparison to undoped samples other than the fact that the dopants seem to have basically no influence on the overall dependence of conductivity on temperature, yielding practically similar activation energies and stagnation of the conductivity increase at temperatures above 500 °C.

For comparison, using the porosity values (cf. Chapter 2, App. A) for these samples, the measured conductivity can be corrected mathematically to represent completely dense samples by means of eq. 4.1 [11]

$$\sigma_{\mathrm{d}} = \frac{\sigma_{\varepsilon}}{1 - \frac{3}{2} \cdot \varepsilon}, \tag{4.1}$$

where σ_{d} denotes the conductivity of a completely dense sample and σ_{ε} the conductivity of a porous sample with the porosity ε.

Fig. 4.6 shows the mean conductivity values for all compositions at the three temperatures of 700 °C, 800 °C and slightly below 900 °C (symbols with error bars) as well as the calculated theoretical conductivities for completely dense samples.

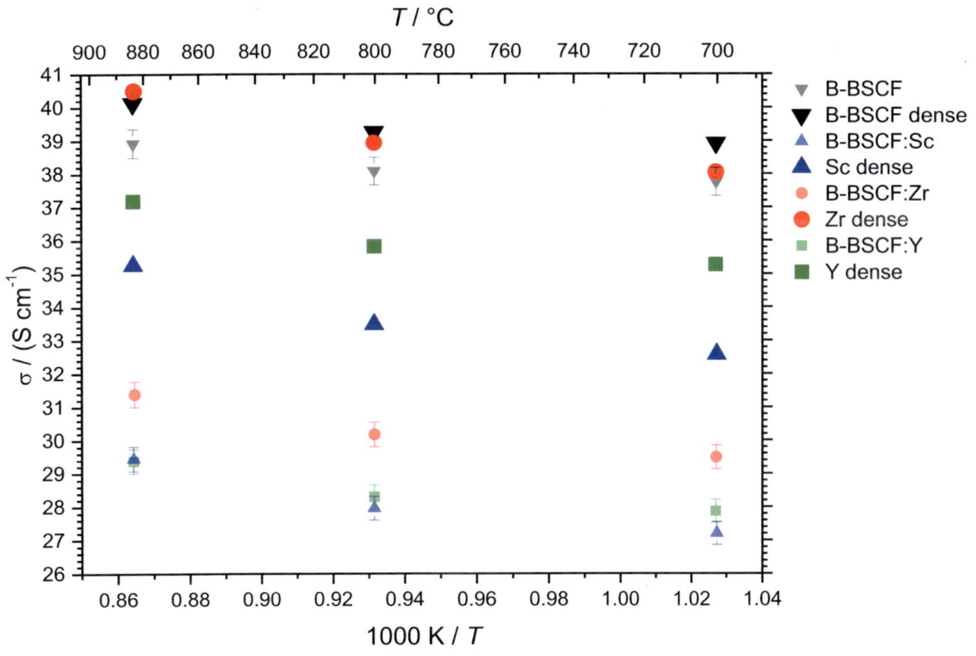

Fig. 4.6: Mean values of the conductivity of doped BSCF samples during cooling in air for three BSCF samples (black), and two samples each for BSCF3Y (green), BSCF3Zr (red) and BSCF3Sc (blue) as well as the calculated corrected values for 100 % dense samples (solid symbols).

It can be seen that generally, doping the BSCF system with another cation reduces the electrical conductivity in comparison to the pure BSCF material. This is the case for the dopants Sc, where conductivity decreases by more than 15% as well as for Y, where the decrease amounts to nearly 10 %. Only the sample containing Zr yields basically comparable conductivity values to the pure system, solidifying the assumption of Chapter 2 that this dopant was not introduced in sizable amounts into the BSCF lattice but rather formed secondary phases.

However, the shown decrease in conductivity for the Sc and especially the Y doped samples does not necessarily lead to a significant drop in oxygen permeation but – in the latter case – yields a significant stabilization of the cubic BSCF phase, as shown in Chapter 8.

4.3.5 Equilibrium Conductivity Values at Intermediate and Low Temperatures

In the earlier parts of this chapter, conductivity curves have always been derived from cool-down measurements. As conductivity changes stem from thermodynamically changing oxygen stoichiometries, the equilibrium values are reached in the order of seconds or minutes at high temperatures above 600 °C, where oxygen exchange is reasonably fast. At lower temperatures, however, the mechanisms of oxygen in- or excorporation slow down significantly, leading to quite long equilibration times of up to several tens of hours. This means that in temperature sweep measurements, the conductivity will show hysteresis and the true equilibrium values below 600 °C can only be determined by suitable dwelling times at chosen temperatures. Fig. 4.7 shows subsequent heating and cooling sweeps for a 3 at% Zr doped BSCF sample including different dwelling times during heating (3rd and 4th measurements – blue and black curves) as well as during cooling (5th and 6th measurements – red and gold curves).

It can be seen that while above temperatures of 650 °C both heating and cooling curves yield the same and therefore the equilibrium conductivity values, at lower temperatures down to 500 °C the true equilibrium values can be reached only by holding the sample at these temperatures for 10 to 20 hours. The fact that the values from dwelling while heating the sample up and those derived from dwelling while cooling down are identical – marked by red squares in Fig. 4.7 – and range between the high- and low-end values derived from the sweeping measurements while heating and cooling without dwell times at chosen temperatures, serves as indicator that equilibrium has indeed been reached.

At even lower temperatures below 500 °C, dwell times of 20 to 50 hours turned out to still be insufficient in order to reach the equilibrium values, marked in Fig. 4.7 by the red shaded ellipsis.

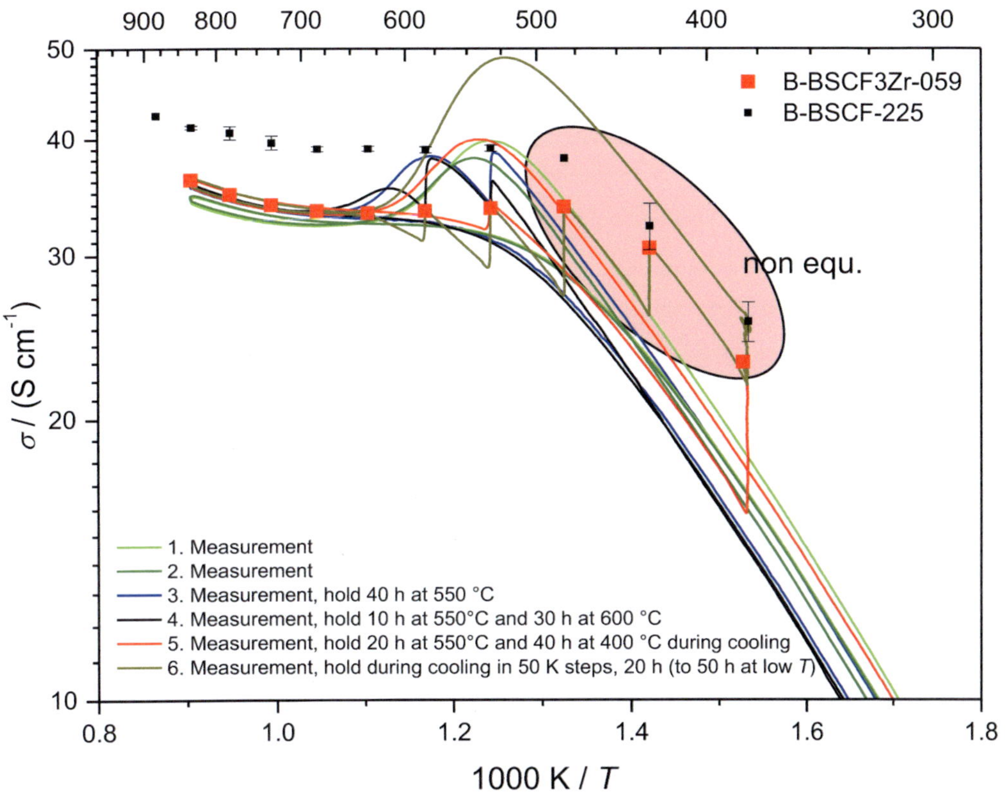

Fig. 4.7: Heating and cooling conductivity curves for a BSCF3Zr sample with different dwelling times at intermediate temperatures, resulting in true equilibrium values above 500 °C (red symbols) in comparison with equally derived equilibrium values for pure BSCF (black symbols).

4.4 Conclusions

The electrical properties of BSCF bulk samples and epitaxial PLD thin films have been measured in pure oxygen and in air atmosphere in the temperature range from 40 to 900 °C. Additionally, MOD-derived polycrystalline thin films have been measured in air for comparison. Dense samples of BSCF doped with 3 at-% Zr, Y, or Sc have been compared to the pure BSCF system. The real equilibrium conductivity values at intermediate temperatures could be determined by long dwell-times.

- BSCF bulk conductivity in pure oxygen and air atmospheres exhibits quite similar activation energies of 0.361 eV and 0.358 eV, respectively.

- PLD thin film conductivity in air is two to five times higher in the whole temperature range due to the absence of grain boundaries and substrate induced stress.
- Similar behavior for the PLD thin films in comparison to bulk samples is found at low temperatures below approx. 400 °C, while at higher temperatures conductivity continues increasing, possibly related to the stress induced by the substrate and/or to the crystallographic orientation of the surface planes.
- PLD thin films in pure oxygen and air atmospheres show similar conductivity behavior with two temperature ranges showing different activation energies of 0.243 eV from 40-300 °C and 0.168 eV from 300-900 °C in pure oxygen and 0.254 eV and 0.182 eV in air in the respective temperature ranges.
- For polycrystalline MOD thin films, three different regimes become visible in the conductivity measurements for temperatures above 400 °C, between 350 °C and 250 °C as well as below 250 °C. This could be explained by the presence of grain boundaries while still being subject to substrate induced stress.
- Dopants have basically no influence on the overall dependence of conductivity on temperature, yielding practically similar activation energies and stagnation of the conductivity increase at temperatures over 500 °C, while generally reducing the conductivity in comparison to pure BSCF.
- At temperatures below 600 °C, equilibrium conductivity values can only be determined by relatively long dwelling times of several tens of hours and have been calculated for pure BSCF as well as BSCF3Zr.

References

[1] M. Burriel, C. Niedrig, W. Menesklou, S. F. Wagner, J. Santiso and E. Ivers-Tiffée, "BSCF epitaxial thin films: Electrical transport and oxygen surface exchange", *Solid State Ionics* **181**, pp. 602-608 (2010).

[2] K. Asano, C. Niedrig, W. Menesklou, S. F. Wagner and E. Ivers-Tiffée, "Modification of Oxygen / $(Ba_{0.5}Sr_{0.5})(Co_{0.8}Fe_{0.2})O_{3-\delta}$ Interfaces Derived By Metal-Organic Deposition", *ECS Transactions* **66**, pp. 147-155 (2015).

[3] H. J. M. Bouwmeester, H. Kruidhof and A. J. Burggraaf, "Importance of the surface exchange kinetics as rate limiting step in oxygen permeation through mixed-conducting oxides", *Solid State Ionics* **72**, pp. 185-194 (1994).

[4] H. Middleton, S. Diethelm, R. Ihringer, D. Larrain, J. Sfeir and J. van Herle, "Co-casting and co-sintering of porous MgO support plates with thin dense perovskite layers of $LaSrFeCo_3$", *Journal of the European Ceramic Society* **24**, pp. 1083-1086 (2004).

[5] L. Hong, X. Chen and Z. Cao, "Preparation of a perovskite $La_{0.2}Sr_{0.8}CoO_{3-x}$ membrane on a porous MgO substrate", *Journal of the European Ceramic Society* **21**, pp. 2207-2215 (2001).

[6] C. Li, T. Hu, H. Zhang, Y. Chen, J. Jin and N. Yang, "Preparation and characterization of supported dense oxygen permeating membrane of mixed conductor $La_2NiO_{4+\delta}$", *Journal of Membrane Science* **226**, pp. 1-7 (2003).

[7] H. J. M. Bouwmeester and A. J. Burggraaf, "Dense Ceramic Membranes for Oxygen Separation", in P. J. Gellings and H. J. M. Bouwmeester (Eds.), *The CRC Handbook of Solid State Electrochemistry*, Boca Raton, FL: CRC Press, pp. 481-553 (1997).

[8] W. Zhou, R. Ran, Z. P. Shao, W. Zhuang, J. Jia, H. X. Gu, W. Q. Jin and N. P. Xu, "Barium- and strontium-enriched $(Ba_{0.5}Sr_{0.5})_{(1+x)}Co_{0.8}Fe_{0.2}O_{3-\delta}$ oxides as high-performance cathodes for intermediate-temperature solid-oxide fuel cells", *Acta Materialia* **56**, pp. 2687-2698 (2008).

[9] B. Wei, Z. Lu, X. Q. Huang, J. P. Miao, X. Q. Sha, X. S. Xin and W. H. Su, "Crystal structure, thermal expansion and electrical conductivity of perovskite oxides $Ba_xSr_{1-x}Co_{0.8}Fe_{0.2}O_{3-\delta}$ ($0.3 \leq x \leq 0.7$)", *Journal of the European Ceramic Society* **26**, pp. 2827-2832 (2006).

[10] Z. H. Chen, R. Ran, W. Zhou, Z. P. Shao and S. M. Liu, "Assessment of $Ba_{0.5}Sr_{0.5}Co_{1-y}Fe_yO_{3-\delta}$ (y=0.0-1.0) for prospective application as cathode for IT-SOFCs or oxygen permeating membrane", *Electrochimica Acta* **52**, pp. 7343-7351 (2007).

[11] A. K. Tagantsev, V. O. Sherman, K. F. Astafiev, J. Venkatesh and N. Setter, "Ferroelectric Materials for Microwave Tunable Applications", *Journal of Electroceramics* **11**, pp. 5-66 (2003).

Chapter 5

Conductivity of MIEC at different pO_2

5.1 Introduction

In Chapter 4, the electrical conductivity behavior of dense BSCF samples, dense samples of doped BSCF, epitaxial BSCF thin films produced by PLD and polycrystalline MOD derived BSCF thin films was examined and discussed depending on temperature. The measurements shown in that chapter also give a glimpse at different conductivity values in different atmospheres, meaning different ambient pO_2 values. In this chapter the electrical conductivity dependence on oxygen partial pressure is examined, providing information about the variation of the electronic transport behavior from the feed side to the permeation side of an oxygen separation membrane.

5.2 Experimental

Contacting of the samples shown in this chapter has been done similar to the ones shown in Chapter 4: for the conductivity measurements two parallel gold contacts were painted at the surfaces of the epitaxial thin-film samples and fired for one hour at a temperature of 950 °C in air. The MOD films were contacted with two parallel gold contacts by magnetron sputtering. Owing to the fairly high electrical resistance of the thin films, their planar resistance was measured in two-point-technique in the oxygen pump setup (cf. Chapters 2, 3) where the inner oxygen partial pressure can be

Parts of the results shown in this chapter have been published in [1;2]

adjusted between 10^{-18} bar and 1 bar within a temperature range of approx. 700-1000 °C by application of a Nernst voltage [3]. The dense bulk samples were measured in four-point-technique in the same setup, electrical contacts were applied by means of Pt wires and paste fired for one hour at 1050 °C in air for the samples shown in 5.3.1 and 5.3.2. All samples shown in 5.3.3 have been contacted the same way after first applying Pt contacts by magnetron sputtering.

5.3 Results and Discussion

5.3.1 Dense BSCF Samples and Epitaxial BSCF Thin Films

In Fig. 5.1 the oxygen partial pressure and related total conductivity changes of a dense BSCF bulk sample are shown as a function of time for five different temperatures as an example of a typical measurement in the oxygen pump setup.

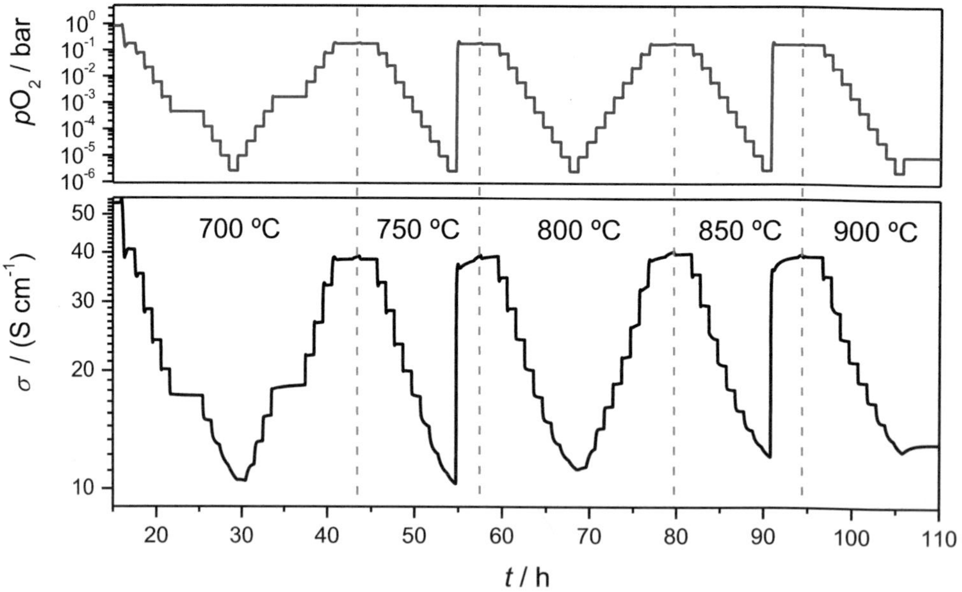

Fig. 5.1: Typical conductivity measurement in the oxygen pump setup for a BSCF sample (FZJ-BSCF-02). The electrical conductivity (black curve) is measured while changing pO_2 in steps (blue curve) at different temperatures.

Every oxygen partial pressure change is followed by a 1-hour dwelling period, allowing for the oxygen exchange to take place until equilibrium is reached. Subsequently, the next partial pressure change is performed. The measurements were taken in steps by first decreasing the oxygen partial pressure from pure oxygen to a pO_2 of approx. $3 \cdot 10^{-6}$ bar followed by reversely increasing the oxygen partial pressure to the atmospheric pressure of 0.21 bar again – using the same oxygen partial pressure values for temperatures of 700 °C, 800 °C and 900 °C, while just jumping back to ambient pressure at the intermediate temperatures of 750 °C and 850 °C.

At 700 °C, there is a longer dwelling time at two pO_2 steps around ~10^{-3} bar, revealing the conductivity response of the sample to consist of two different involved processes, the second one being due to an ongoing phase transition in the BSCF material, discussed in detail in Chapter 8. This phase change, however, does not greatly affect the results shown in the following, as the measurement times of the shown experiments only amount to ~80 hours at temperatures below 850 °C. This time frame is a lot shorter than the needed equilibration time for the happening phase changes – that amount to more than 800 hours in these bulk samples (cf. Chapter 8). Possible deviations caused by the beginning phase changes will amount to less than 1 % as can be seen in Fig. 8.8 in Chapter 8.

For oxygen partial pressures below 10^{-4} bar, the conductivity value did not reach equilibrium after 1 hour and continued decreasing with time. This seems to be due to the rapidly decreasing oxygen exchange parameters – namely the chemical diffusion coefficient D^{δ} and the surface exchange coefficient k^{δ} – at these low oxygen partial pressures, as is discussed in detail in Chapter 6. Equilibrium conductivity values for lower oxygen partial pressures can thus only be obtained through much longer dwelling times, as shown later in this chapter (Figs. 5.6, 5.8). Additionally, at temperatures below 850 °C, the conductivity response to the oxygen partial pressure change always seems to consist of two different processes – as revealed by the longer dwelling times in the 700 °C measurement – first a fast oxygen exchange, followed by a second non-reversible process (cf. Fig. 8.1, Ch. 8), representing a phase transition in the BSCF matrix as discussed in Chapter 8.

A similar measurement was performed on an epitaxial BSCF thin film on NGO with a thickness of 340 nm. In Fig. 5.2 the conductivity values of both the bulk sample and the thin film which had been reached after 1 hour equilibration are plotted versus the oxygen partial pressure in a double logarithmic plot. Values below 10^{-4} bar are not equilibrium values, as the dwelling time was too short to reach oxygen equilibration in both samples.

Conductivity values of the thin film are several times higher than those of the dense bulk sample, due to the intrinsic properties of the thin film/substrate system (cf. Chapter 4). At 900 °C the variation of the conductivity with pO_2 is similar for both the thin film sample and the dense bulk sample, exhibiting a slope of $\frac{d\log(\sigma/(\mathrm{S/cm}))}{d\log(pO_2/\mathrm{bar})} = 0.10$. It seems the high temperature enables the thin film to emit a corresponding amount of oxygen in comparison to the bulk sample due to partially being able to overcome the stress introduced by the NGO substrate with its smaller lattice constant and therefore showing a conductivity change similar to the behavior of the dense bulk.

At 700 °C however, the thin film's conductivity proves much less dependent on the ambient pO_2, showing a slope of only 0.06, probably due to the restrictions on BSCF lattice expansion – coinciding with oxygen loss in the lattice – applied by the substrate.

For the bulk sample, at this lower temperature the case is reversed, its conductivity shows a stronger dependency on oxygen partial pressure than at 900 °C, expressed in a steeper slope of 0.15. In comparison to the thin film, this is understandable as the bulk is relatively unconstrained in respect to lattice parameter growth and therefore able to release more oxygen than the constrained thin film.

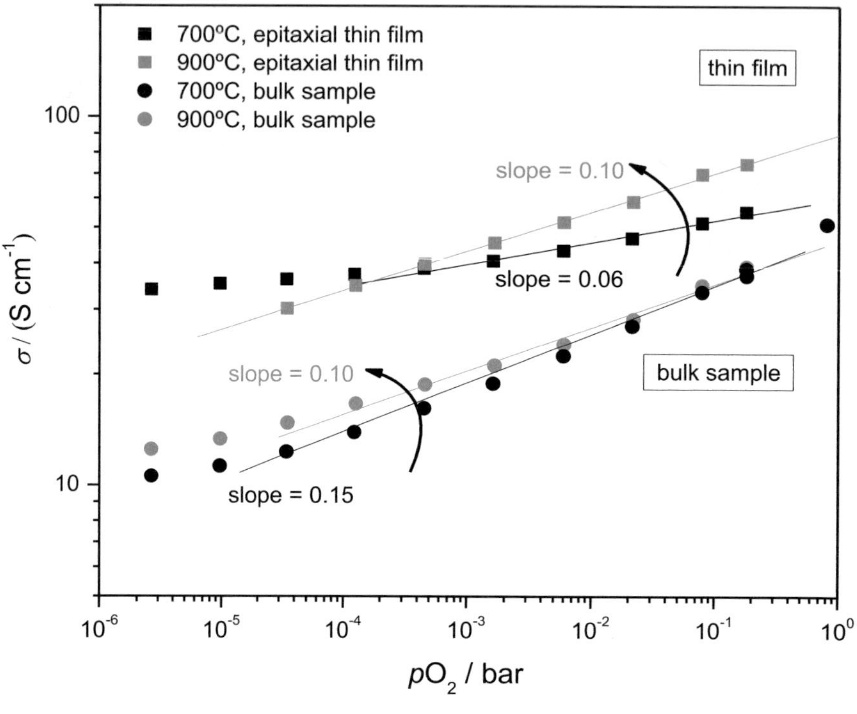

Fig. 5.2: Comparison of the equilibrium conductivity values (above 10^{-4} bar pO_2) of a dense BSCF sample (FZJ-BSCF-02, circles) and an epitaxial BSCF thin film sample (PLD-BSCF-17n2, squares) at 900 °C (red symbols) and 700 °C (black symbols).

In comparison to the bulk's behavior at 900 °C however, the explanation must be found in the conductivity behavior corresponding to temperature shown in Chapter 4. The BSCF bulk sample's conductivity at both temperatures lies in the range of nearly constant conductivity values above approx. 500 °C (cf. Fig. 4.3 in Chapter 4), meaning the loss of positive holes as charge carriers while emitting oxygen is approximately compensated by increased charge carrier mobility when the temperature is increased. At 900 °C, at the lowest oxygen concentration in the lattice, especially at lower oxygen partial pressures of 10^{-3} to 10^{-4} bar, this charge carrier mobility increase seems to be visible in the overall higher conductivity and therefore overshadowing the effect of oxygen loss, further discussed in the following section.

5.3.2 Comparison with PSCF, LSCF and LSC

Similar measurements have been conducted on dense bulk samples of PSCF, LSCF and LSC. Due to the low values of its oxygen exchange parameters (cf. Fig. 6.5, Chapter 6), PSCF only yields equilibrium values for temperatures of 900 °C and above, while for the other compositions stable values could be extracted at 700, 800 and 900 °C. Figs. 5.3 and 5.4 show the results for all compositions, the marked data points at low pO_2 values are non-equilibrium values, as the dwelling time was too short (see previous part of this chapter). Linear fits of the stable equilibrium values are used to evaluate the grade of conductivity dependency for the different compositions.

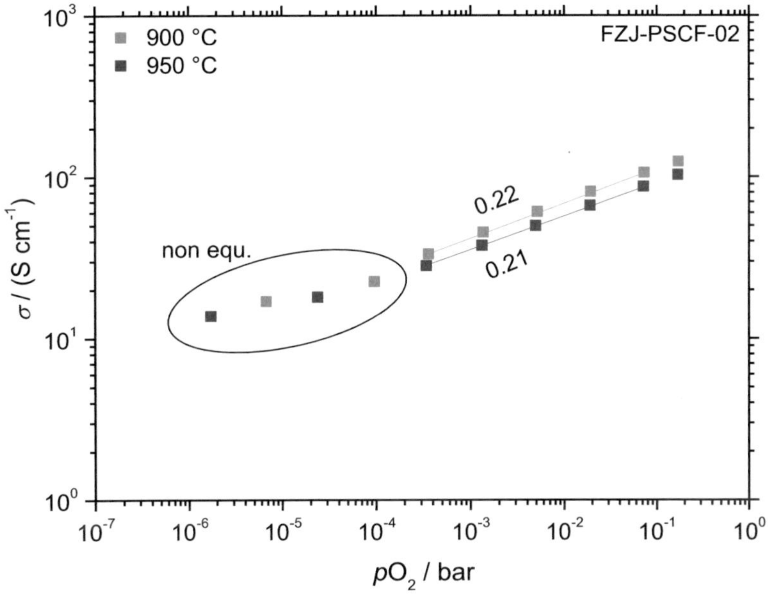

Fig. 5.3: Equilibrium conductivity values at 900 °C (red symbols) and 950 °C (blue symbols) – where obtainable – for a dense PSCF sample (FZJ-PSCF-02).

PSCF (Fig. 5.3) exhibits nearly the same slope of 0.22 and 0.21 at temperatures of 900 °C and 950 °C, respectively. This was to be expected due to the little difference in temperatures. This slope is in good agreement with that of other p-type conductors, e.g., $SrTiO_3$, under oxidizing conditions [4]. The absolute values of conductivity in air match those reported by Kostogloudis and Ftikos [5] very well, pO_2-dependent data are not available in literature.

LSCF also performs with a slope of 0.21 at 900 °C, similar to PSCF, while at lower temperatures of 800 °C and 700 °C the slope decreases to 0.18 and 0.12, respectively (cf. Fig. 5.4, top). At 700 °C, LSCF therefore behaves nearly similar to BSCF with its slope of 0.15. The absolute values of conductivity (in air) are in good agreement with data reported by Kostogloudis and Ftikos [6], while slightly lower than those reported by Bouwmeester et al. [7] who also state a concave

$\sigma(pO_2)$ behavior, attributed to a similar behavior of both $k^\delta(pO_2)$ and $D^\delta(pO_2)$, tentatively due to an ordering of oxygen vacancies at lower pO_2. A slightly concave behavior can also be suggested for the data points shown in Fig. 5.4 (top), contrary to all other three compositions. For the sake of comparison, however, a pseudolinear fit has been performed.

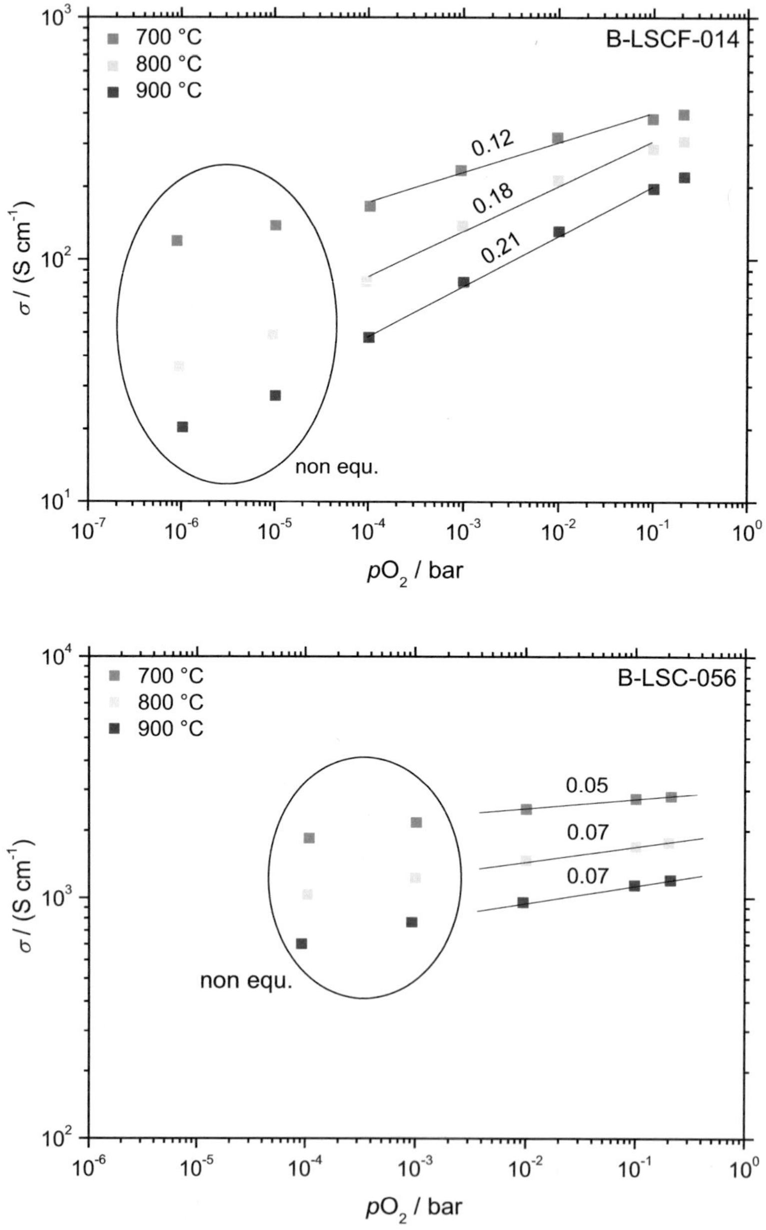

Fig. 5.4: Equilibrium conductivity values at different temperatures between 700 °C and 900 °C – where obtainable – for dense samples of LSCF (B-LSCF-014, top) and LSC (B-LSC-056, bottom).

In the case of LSC (Fig. 5.4, bottom), the slope only differs from 0.05 for 700 °C to 0.07 for both 800 and 900 °C. The latter value is in exact agreement with Søgaard et al. [8].

A comparison between all compositions is also shown in Fig. 5.5. Contrary to the BSCF sample, for all other compositions lower temperature leads to a decreased dependency of the conductivity on pO_2, coinciding with the steady increase of conductivity with temperature in the case of BSCF as opposed to PSCF, LSCF, and LSC which exhibit decreasing conductivity at temperatures above ~ 500 °C. This conductivity behavior could be explained by highly different magnitudes of nonstoichiometry changes in the two different material groups (BSCF vs. PSCF, LSCF, and LSC) as listed from literature data for a temperature of 800 °C and oxygen partial pressures of 1 bar and 10^{-2} bar in Table 5.1.

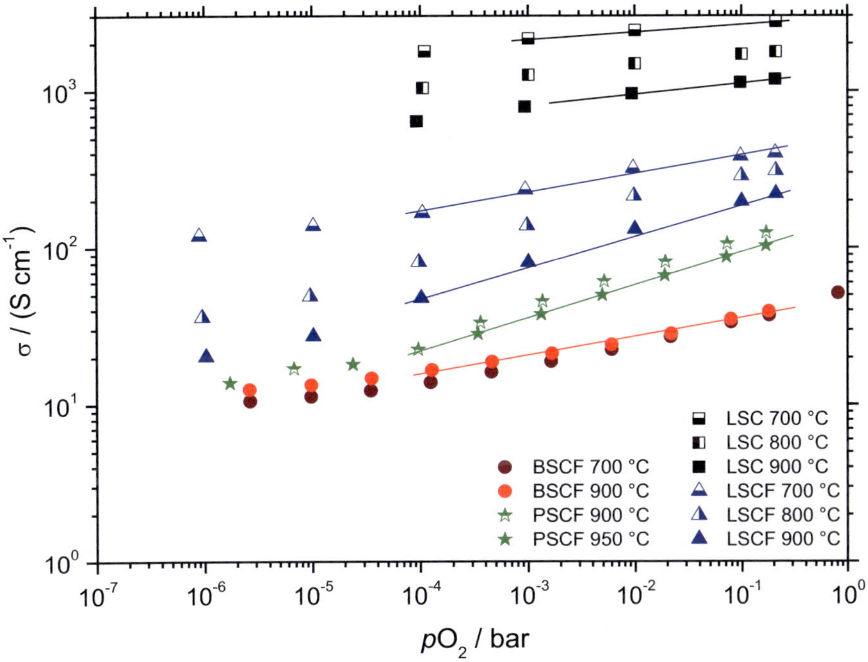

Fig. 5.5: Equilibrium conductivity values at different temperatures between 700 °C and 950 °C (where obtainable) for dense samples of LSC (B-LSC-056, black symbols) LSCF (B-LSCF-014, blue symbols) and PSCF (FZJ-PSCF-02, green symbols) in comparison with BSCF (FZJ-BSCF-02, red symbols).

As nonstoichiometry changes directly affect charge carrier concentration – every oxygen vacancy created is accompanied by an annihilation of two positive charge carriers (holes) – the small relative change $\Delta\delta$ in BSCF (~ 8 % between 1 bar and 10^{-2} bar pO_2 at 800 °C) is overcompensated by an increased charge carrier mobility at high temperatures, resulting in the steady increase of conductivity. For the other compositions, however, charge carrier concentration is approximately halved in the mentioned pO_2 range (LSCF: $\Delta\delta$ > 150 %, LSC: $\Delta\delta$ > 80 %), resulting in a decrease in conductivity after the maximum at intermediate temperatures – at which notable oxygen exchange with the ambient atmosphere sets in. For, e.g., LSCF this can be seen by the general conductivity values for normal air at the three different temperatures in the shown graph (rightmost data points at 0.21 bar pO_2) and also reported, e.g., by [6].

composition	δ at pO_2 = 1 bar	δ at pO_2 = 10^{-2} bar	relative change
BSCF [9]	0.72	0.78	8 %
PSCF	n.a.	n.a.	n.a.
LSCF [7]	0.03	0.08	167 %
LSC [8]	0.07	0.13	86 %

Tab. 5.1: Nonstoichiometry values δ at 800 °C at 1 bar and 10^{-2} bar pO_2.

The small relative stoichiometry change for BSCF is also an indication that BSCF is situated near the edge of the stability range for cubic perovskites, meaning that at steadily increasing temperature the conductivity changes with pO_2 become *less* pronounced (slopes $\frac{d \log \sigma}{d \log pO_2}$ changing from 0.15 at 700 °C to 0.10 at 900 °C), while in the other materials – where very high relative changes in oxygen nonstoichiometry can safely be achieved – increasing temperature leads to *more* pronounced conductivity changes with pO_2 (slopes for LSCF change from 0.12 at 700 °C to 0.21 at 900 °C, slopes for LSC change from 0.05 at 700 °C to 0.07 at 900 °C).

5.3.3 Comparison of BSCF with Zr-, Y-, and Sc-doped BSCF

As the doping of BSCF with different cations can enhance the stability of the cubic matrix at temperatures of interest (cf. Chapter 8), the same measurements have been conducted on dense samples of BSCF3Zr, BSCF3Y, and BSCF3Sc. Fig. 5.6 shows a comparison of the equilibrium conductivity values at different oxygen partial pressures between the three different compositions and undoped BSCF at 900 °C. All conductivity values have been corrected for the different sample porosities as described in Chapter 4.

The highest conductivity at all oxygen partial pressures belongs to the undoped BSCF, followed by the Zr-,Y- and finally Sc-doped material. This is consistent with the findings at ambient air pressure in Chapter 4. Unsurprisingly, the dependency on pO_2 is nearly similar for all compositions, as this small amount of B-site cation substitution does not affect oxygen transport properties on a large scale.

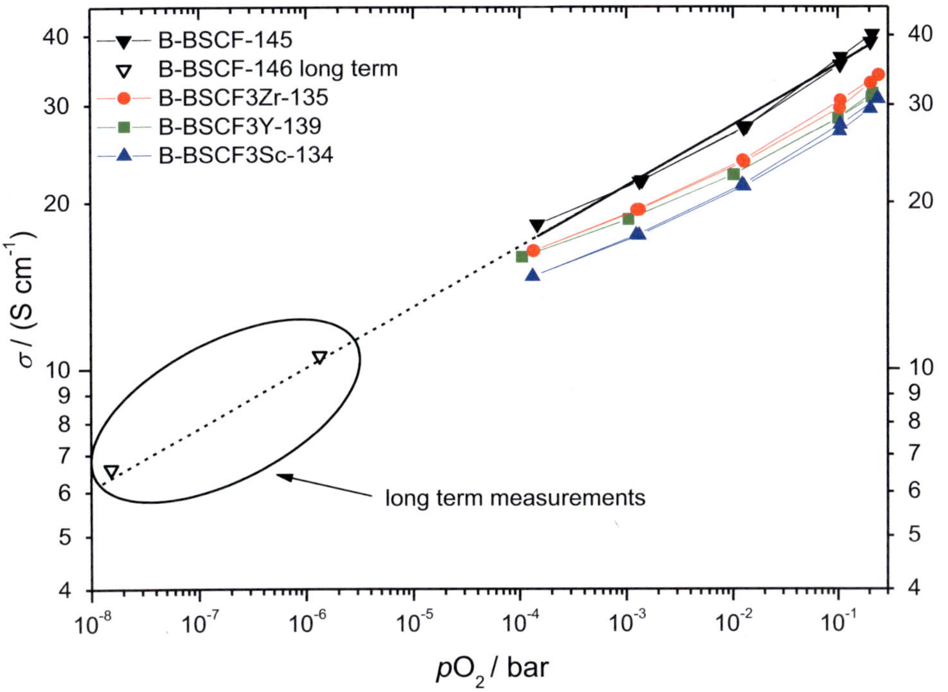

Fig. 5.6: Equilibrium conductivity values at 900 °C for BSCF3Zr (red symbols), BSCF3Y (green) and BSCF3Sc (blue) in comparison to undoped BSCF (black symbols, long term measurements (hollow symbols) conducted on an identical BSCF bulk sample: B-BSCF-146). Conductivity has been corrected for sample porosity.

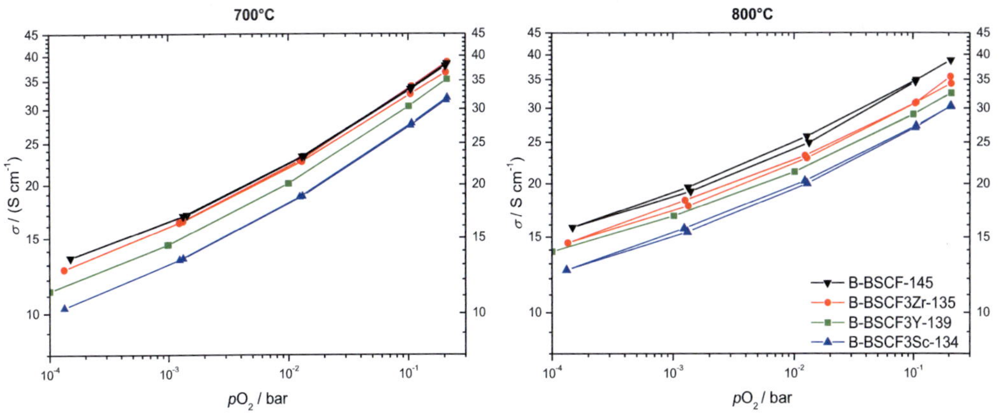

Fig. 5.7: Equilibrium conductivity values at 700 °C (left) and 800 °C (right) for BSCF3Zr (red symbols), BSCF3Y (green symbols) and BSCF3Sc (blue symbols) in comparison to undoped BSCF (black symbols).

Equivalent results at lower temperatures are shown in Fig. 5.7 for 700 °C (left) and 800 °C (right).

The two marked data points in Fig. 5.6 are equilibrium values for an identical BSCF bulk sample derived from two long term measurements at these low oxygen partial pressures of approx. 10^{-6} and 10^{-8} bar. As indicated by the dashed line, they fit very well to the linear fit (solid black line) for all data points above 10^{-4} bar.

Fig. 5.8 shows the long term measurement for the equilibrium conductivity value at a pO_2 near 10^{-6} bar. Reaching a stable value required more than 600 hours. The red squares indicate the starting and end values of the conductivity. The starting value was measured when the oxygen partial pressure in the sample space reached 0.21 bar after flushing with pure oxygen and starting the electrochemical pumping, while the end value was taken when a stable conductivity was reached after pumping back from the low pressure to 0.21 bar. They amount to 33.0 S cm^{-1} and 32.99 S cm^{-1} respectively, indicating that the long equilibration is not due to some kind of deterioration in the sample – i.e. phase transformations or decomposition – but in fact an indicator of the strongly decreasing values for the transport parameters at low pO_2.

The similar measurement to obtain the equilibrium value at approx. 10^{-8} bar pO_2 shown in Fig. 5.6 took an even longer equilibration time of 800 hours, while also showing reversibility.

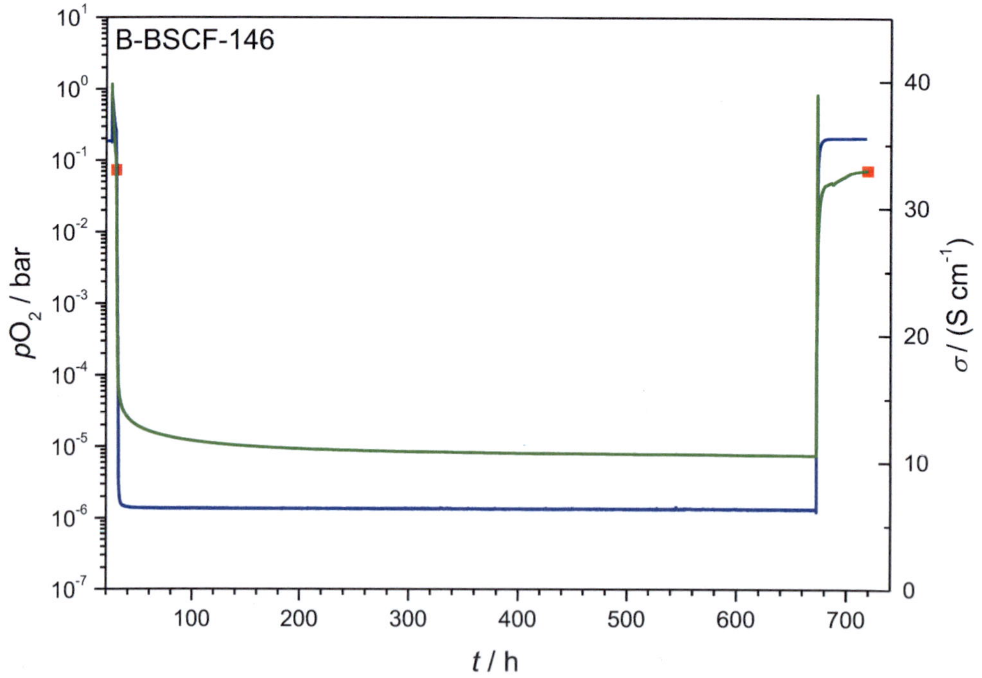

Fig. 5.8: Long term conductivity measurement at 900 °C and 10^{-6} bar pO_2 for a BSCF sample (B-BSCF-146). Equilibration takes more than 600 hours, while the practically identical starting and end values at 0.21 bar pO_2 – red squares – show the reversibility of the process.

5.4 Conclusions

In this chapter, the "oxygen pump" setup was used to measure the dependency of the electrical conductivity of bulk samples of different MIEC materials as well as an epitaxial BSCF thin film on oxygen partial pressure at temperatures from 700 °C to 900 °C

- Equilibrium values at oxygen partial pressures below 10^{-4} bar (10^{-2} bar in the case of LSC) cannot be reached within the dwelling time of 1 hour even at the highest temperature, due to the rapidly decreasing oxygen exchange parameters D^{δ} and k^{δ} at these low oxygen partial pressures.

For the pure BSCF system, bulk samples have been compared to epitaxial thin films:

- For a BSCF bulk sample, equilibration at 10^{-6} and 10^{-8} bar pO_2 can be achieved at 900 °C after a dwelling time of several hundred hours, fitting in well with the values obtained at higher pO_2. Reversibility has been shown.
- Conductivity dependency on pO_2 behaves similar for BSCF bulk samples and epitaxial thin films at 900 °C, exhibiting a slope of 0.10 both, while the thin films' dependency lessens at lower temperatures of 700 °C – with a slope of 0.06 in relation to 0.15 for the bulk sample.

Dense bulk samples of PSCF, LSCF, and LSC have also been measured for comparison:

- PSCF only yields equilibrium values for temperatures of 900 °C and above due to the low values of its oxygen exchange parameters. It shows nearly the same slope of 0.22 and 0.21 at temperatures of 900 °C and 950 °C, respectively.
- LSCF also performs with a slope of 0.21 at 900 °C, similar to PSCF, while at lower temperatures of 800 °C and 700 °C the slope decreases to 0.18 and 0.12, respectively. The latter value is nearly similar to the 0.15 for BSCF.
- For LSC, the slope only differs from 0.05 for 700 °C to 0.07 for both 800 and 900 °C due to the much smaller range for the oxygen nonstoichiometry in the lattice.

Bulk samples of the proposed doped materials BSCF3Zr, BSCF3Y and BSCF3Sc have been compared to the performance of pure BSCF:

- BSCF shows the highest conductivity values at all partial pressures, followed by the Zr-, Y-, and finally Sc-doped material. This is consistent with the previous findings at ambient air pressure.
- pO_2-dependency is nearly similar for all compositions, as this small amount of B-site cation substitution presumably does not affect oxygen transport properties on a large scale.

References

[1] C. Niedrig, S. F. Wagner, W. Menesklou and E. Ivers-Tiffée, "Characterization of oxygen-dependent stability of selected mixed-conducting perovskite oxides", *Solid State Ionics* **273**, pp. 41-45 (2015).

[2] E. Ivers-Tiffée, C. Niedrig and S. F. Wagner, "MIEC Materials for Membrane Applications: Enhancing the Oxygen Transport", *ECS Transactions* **61**, pp. 283-293 (2014).

[3] K. Beetz, *Die geschlossene Festelektrolyt-Sauerstoffpumpe*, Düsseldorf: VDI Verlag GmbH (1993).

[4] W. Menesklou, H.-J. Schreiner, K. H. Härdtl and E. Ivers-Tiffée, "High temperature oxygen sensors based on doped $SrTiO_3$", *Sensors and Actuators B-Chemical* **59**, pp. 184-189 (1999).

[5] G. C. Kostogloudis and C. Ftikos, "Oxygen nonstoichiometry in $Pr_{1-x}Sr_xCo_{0.2}B_{0.8}O_{3-\delta}$ (B = Mn, Fe, x = 0.2, 0.4) perovskite oxides", *Journal of the European Ceramic Society* **27**, pp. 273-277 (2007).

[6] G. C. Kostogloudis and C. Ftikos, "Properties of A-site-deficient $La_{0.6}Sr_{0.4}Co_{0.2}Fe_{0.8}O_{3-\delta}$-based perovskite oxides", *Solid State Ionics* **126**, pp. 143-151 (1999).

[7] H. J. M. Bouwmeester, M. W. den Otter and B. A. Boukamp, "Oxygen transport in $La_{0.6}Sr_{0.4}Co_{1-y}Fe_yO_{3-\delta}$", *Journal of Solid State Electrochemistry* **8**, pp. 599-605 (2004).

[8] M. Søgaard, P. V. Hendriksen, M. Mogensen, F. W. Poulsen and E. Skou, "Oxygen nonstoichiometry and transport properties of strontium substituted lanthanum cobaltite", *Solid State Ionics* **177**, pp. 3285-3296 (2006).

[9] S. McIntosh, J. F. Vente, W. G. Haije, D. H. A. Blank and H. J. M. Bouwmeester, "Oxygen stoichiometry and chemical expansion of $Ba_{0.5}Sr_{0.5}Co_{0.8}Fe_{0.2}O_{3-\delta}$ measured by in situ neutron diffraction", *Chemistry of Materials* **18**, pp. 2187-2193 (2006).

Chapter 6

Conductivity Relaxation Measurements in the Oxygen Pump

6.1 Introduction

For the different applications, oxygen partial pressures at the surfaces of the mixed ionic electronic conducting (MIEC) material may typically range between, e.g., pure oxygen atmospheres and values as low as $pO_2 = 10^{-4}$ bar: oxygen transport membranes (OTMs) operated in four-end mode, i.e. with pressurized air on the feed side and flue gas recirculation on the permeate side, may encounter low-end values of $pO_2 \approx 50$ mbar, in three-end mode the vacuum pump on the permeate side may even lead to slightly lower pO_2 values of around 1 mbar. In SOFC application, a cathode overpotential of up to 200 mV can result in partial pressures as low as 10^{-4} bar (at 750 °C) at the MIEC surface.

Therefore, an investigation of the oxygen transport kinetics – namely the chemical diffusion coefficient D^{δ} and the surface exchange coefficient k^{δ} – is of interest not only at the oxygen partial pressure of ambient air as discussed on thin films in the next chapter, but for a much broader pO_2 range as these values determine the performance of the materials.

The kinetics of oxygen transport can be studied using electrical conductivity relaxation (ECR) in the time domain – as opposed to the frequency domain approach presented in the next chapter. This method involves monitoring of the transient conductivity exhibited by the MIEC, at fixed temperature, after an instantaneous change of the ambient pO_2. The chemical diffusion coefficient D^{δ} and the surface exchange coefficient k^{δ} can then be obtained from a single experiment by fit-

Parts of the results shown in this chapter have been published in [1-6].

ting the transient conductivity to the appropriate solution of Fick's second law [7] for given boundary conditions (determined by the sample dimensions and the setup characteristics).

Commonly, the instantaneous pO_2 change in such an experiment is achieved by switching between two different gas streams with a different pO_2. In this chapter, the oxygen transport parameters of BSCF, LSCF and PSCF are studied by means of ECR as a function of temperature and pO_2 in the respective ranges $700 \leq T\,/\,°C \leq 900$ and $10^{-6} \leq pO_2\,/\,bar \leq 0.21$. Rather than using different gas streams with known pO_2, the oxygen coulometric titration setup shown earlier (cf. Chapters 2, 3) is used to control the pO_2 at the sample position and to perform step changes in the pO_2. For BSCF, LSCF, and PSCF, k^{δ} values could be determined for the first time at 900 °C as a function of pO_2.

6.2 Experimental

6.2.1 Contacting of the Samples

For conductivity measurements all bulk samples were contacted with Pt wires using a frit-free Pt paste fired at 1050 °C for 1 h in ambient air. Owing to the fairly low electrical resistance of the MIEC bulk samples, electrical measurements were all carried out in 4-point technique. The bar-shaped dense bulk samples had the following geometries: $10 \cdot 6 \cdot 1.2$ mm³ (inner contact distance: 3.8 mm) for the BSCF sample, $10 \cdot 6 \cdot 1.1$ mm³ (inner contact distance: 6.6 mm) for the LSCF sample, and $10 \cdot 6 \cdot 1.1$ mm³ (inner contact distance: 4.4 mm) for the PSCF sample.

6.2.2 Measurements

Conductivity relaxation measurements were performed on the dense MIEC bulk samples at pO_2 values in the range between 10^{-6} and 0.21 bar in not-too-large pO_2 steps (thus remaining close to chemical equilibrium) at temperatures between 700 °C and 900 °C in order to determine D^{δ} and k^{δ} values as a function of temperature and pO_2.

One limitation of the setup is given by the speed of the oxygen pumping through the zirconia at around 10^{-1} bar pO_2 and above. Under such highly oxidizing conditions one-decade step changes in pO_2 may take up to several minutes, as the maximum amount of oxygen transported through the YSZ tube is limited by the inherent pumping capabilities of the setup (cf. [1], Chapters 2, 3). Regarding sample oxygen exchange kinetics, these pumping times might be too long for parameter extraction. Despite this limitation, however, the setup provides the possibility of precise measurements at oxygen partial pressure ranges not easily accessible using gas mixtures.

6.3 Results and Discussion

6.3.1 Possible Error Sources of ECR Measurements

In the time domain measurements, where an MIEC sample's conductivity response is measured after the ambient oxygen partial pressure pO_2 is abruptly changed using different gas mixtures flowing around the sample or by chemically pumping oxygen into or out of a static atmosphere surrounding the sample, both transport parameters D^δ and k^δ are obtained from a single experiment by fitting the transient conductivity to the appropriate solution of Fick's second law for given boundary conditions (determined by the sample dimensions and the setup characteristics).

In principle, this method is prone to two different potential sources of error, one being the conditions of the sample itself, the other introduced by the method of changing the atmosphere around the sample.

For the sample itself, this amounts to the knowledge of exact sample geometry (for appropriate application of the mathematical solution), surface conditions that influence oxygen surface exchange rates (different morphologies after polishing, presence of potential secondary phases by prolonged contact to trace gases or simply phase transitions in the material itself induced by the measurement conditions) and grain size of the sample (determined by the production process), which influences the relation of grains to grain boundaries at the surface – thus potentially influencing the surface exchange processes – as well as in the whole sample volume – with potential influences on oxygen bulk diffusion.

All these sample specific conditions are also potentially depending on the thermal history of the sample prior to the measurement itself – i.e. the production process of the sample as well as eventual preceding unrelated treatments or measurements. Furthermore, they are prone to change during the measurement itself if the experiment is running for prolonged periods of time and/or at different temperatures.

Invoking conductivity changes in the sample by changing the ambient atmosphere introduces two more problems: by changing the chemical conditions, the sample is per definition out of equilibrium – how much so depends on the step size of the chemical potential change. Therefore, the extracted transport parameters can only be attributed to either the starting conditions prior to the change or the conditions after the step-change (typically, this definition is used). But as at least the surface exchange reaction is influenced by the whole process of the atmospheric change, transport parameters for one specific pO_2 derived from reducing steps usually differ from those derived from an oxidizing change in atmosphere – the amount of discrepancy between both values again depending on the size of the applied steps.

Also, surface exchange values are prone to be different in resting gas atmospheres (pO_2 changes realized by chemically pumping) and in flowing gas (sample space flushed with different gas compositions), as an existing gas flow may influence oxygen adsorption at the sample surface.

Furthermore, changes in gas atmospheres can never be instantaneous, therefore necessitating a mathematical correction for the needed reactor flush time, introducing potential systematic errors.

6.3.2 Conductivity Measurements as $f(pO_2,T)$

For the perovskite-oxide compositions investigated in this work, oxygen exchange with the ambient gas phase at high temperatures occurs according to:

$$\tfrac{1}{2}O_{2,\text{gas}} + V_O^{\bullet\bullet} \leftrightarrow O_O^x + 2h^{\bullet}, \tag{6.1}$$

where O_O^x and $V_O^{\bullet\bullet}$ denote oxygen ions on the regular oxygen lattice sites and oxygen vacancies, respectively (in Kröger-Vink notation), and $h^{\bullet}$ holes. Due to the highly changeable oxygen non-stoichiometry, the dominating (p-type) electronic conductivity is a function of ambient oxygen partial pressure pO_2 (cf. Chapter 5, [2]), as can be seen from Fig. 6.1 for the case of a dense ceramic LSCF sample at three different temperatures (700, 800, and 900 °C). After each oxygen partial pressure change, a period of 4 h was allowed for the equilibration of the sample; after this period of time the next partial pressure change was performed. The measurements were taken in steps of approx. one decade by first decreasing the oxygen partial pressure from a pO_2 of 0.21 bar (after initially flushing the whole system with pure oxygen) down to 10^{-6} bar followed by symmetrical steps increasing the oxygen partial pressure again.

The dashed vertical lines in Fig. 6.1 indicate the times when a temperature increase (from 700 to 800 °C and 800 to 900 °C, respectively) has been performed (with a rate of 1 K min^{-1}). The subsequent changes in conductivity are clearly visible during the 100 min heating time.

In the first run at a temperature of 700 °C there are visible oscillations in the achieved pO_2 at a partial pressure of 10^{-4} bar and below. This demonstrates one difficulty inherent to this measurement setup: Depending on the sample properties and the pO_2 and temperature ranges, the whole system behavior with regard to oxygen pumping is constantly changing and has to be taken into account by adjusting the PID parameters of the controller accordingly for each experimental condition.

It can, however, be seen for all three temperatures that the low-pO_2 (below 10^{-4} bar) oxygen-transport properties of the material are slowed down to such a degree that no equilibrium is reached even after four hours. A similar behavior was observed for the other MIEC materials investigated in this chapter, and even considering dwell times of more than ten hours (not shown here) proved to be insufficient for samples of this geometry. In Chapter 5 it was shown that in the case of BSCF ceramic bulk samples, equilibration at 10^{-6} and 10^{-8} bar pO_2 takes several hundreds of hours.

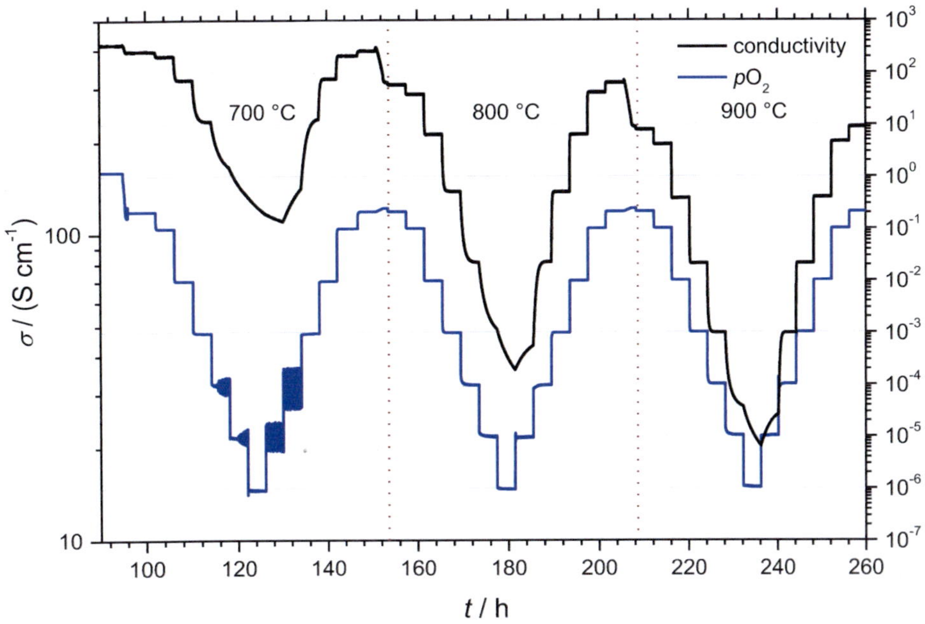

Fig. 6.1: Oxygen partial pressure (blue) and related total conductivity changes (black) of an LSCF bulk sample (B-LSCF-014) shown as a function of time for three different temperatures (700 °C, 800 °C, and 900 °C).

Therefore, transport parameters could not be extracted from conductivity measurements at oxygen partial pressures lower than 10^{-5} bar and the aforementioned oscillations in the low-pO_2 regime could be neglected. Determination of k^{δ} values at even lower pO_2 would require samples with a suitable geometry, e.g. dense thin films, cf. [3]. One should, however, bear in mind that a study of such a purely surface-controlled system cannot yield D^{δ} values; moreover, thin films are much more sensitive to structural changes and stability issues, e.g., formation of the hexagonal phase in the BSCF system (cf. [4;8;9], Chapter 8).

6.3.3 Evaluation of Response Behavior

Fig. 6.2 shows an exemplary oxygen-exchange equilibration process (reduction run) for the LSCF bulk at 800 °C. After changing the pO_2 from 10^{-3} bar down to 10^{-4} bar, thus removing 90 % of the oxygen within the sample chamber – a process that can be seen to take place within a few tens of seconds under such reducing conditions – the relaxation behavior of the sample conductance (shown here as a normalized material conductivity) is nicely visible over a timespan of more than 150 minutes.

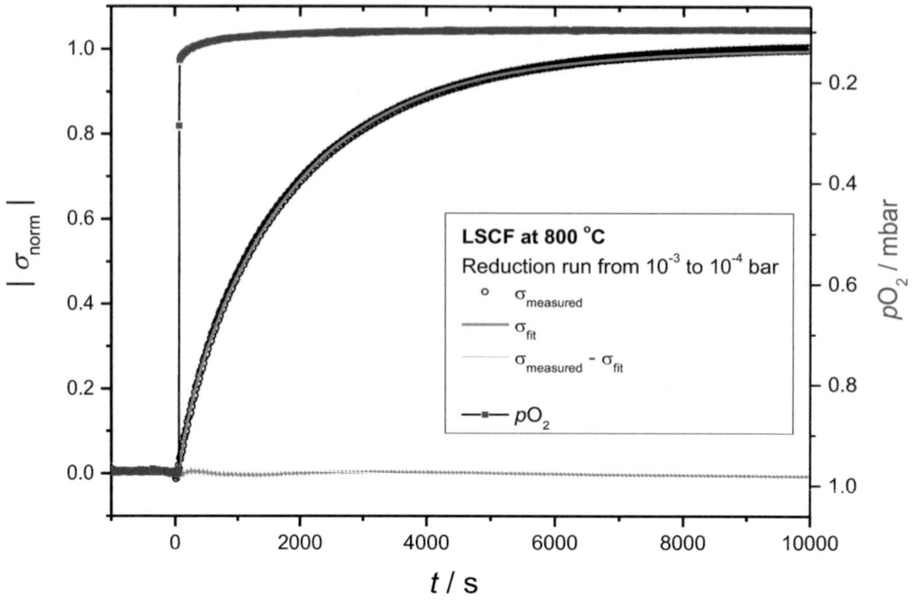

Fig. 6.2: Exemplary reduction run for the LSCF bulk from 10^{-3} bar down to 10^{-4} bar at 800 °C: The measurement data (black open symbols) and the fit function (red curve) show an excellent agreement (the residuals are very close to zero, grey curve). The pO_2 change (blue symbols) takes place within a few tens of seconds, whereas conductivity relaxation occurs over nearly three hours.

Chemical diffusion (D^{δ}) and surface exchange (k^{δ}) coefficients can then be obtained from nonlinear least-squares fits of the solution of the two-dimensional diffusion equation to the experimental relaxation data of the dimensionless transient conductivity as described by den Otter et al. [10;11]:

$$\sigma_{\text{norm}} = \frac{\sigma(t)-\sigma_0}{\sigma_\infty - \sigma_0} = 1-\exp(-\frac{t}{\tau_{\text{f}}}) - \sum_{n=1}^{\infty}\sum_{m=1}^{\infty} A_{n,m} \frac{\tau_{n,m}}{\tau_{n,m}-\tau_{\text{f}}}\left(\exp(-\frac{t}{\tau_{n,m}}) - \exp(-\frac{t}{\tau_{\text{f}}})\right) \tag{6.2}$$

where σ_0 and σ_∞ are the conductivities at time $t = 0$ and $t = \infty$, respectively, τ_{f} is the so-called flushing time of the reactor (the time after which ~63 % of the pO_2 step change is completed [11]), and $2b_i$ is the sample dimension along coordinate i. The time constants are given by

$$\tau_{n,m} = \frac{1}{D^{\delta}\cdot\left[\left(\beta_{x,n}/b_x\right)^2 + \left(\beta_{y,m}/b_y\right)^2\right]} \tag{6.3}$$

and the parameters $\beta_{m,i}$ are evaluated from

$$\beta_{m,i}\tan\beta_{m,i} = L_i = \frac{b_i k^{\delta}}{D^{\delta}} \tag{6.4}$$

Using Eq. (6.2) it is possible to obtain both D^{δ} and k^{δ} from the experimental relaxation data provided that $0.03 < L_i < 30$. As indicated by den Otter et al. [11], D^{δ} cannot be obtained from the relaxation data if $L_i < 0.03$. In that case, the equilibration kinetics is entirely governed by the surface reactions. On the other hand, if $L_i > 30$, the transient is not affected by the surface reactions and only D^{δ} can be obtained from the fitting procedure. Detailed descriptions of the electrical conductivity relaxation technique and the model used for data fitting are given elsewhere [10;11].

The relaxation data shown in Fig. 6.2 demonstrate the good feasibility of this method as long as equilibration takes place within the timescale of the measurement and the relaxation process is solely determined by Eq. (6.2) which may not be the case, e.g., if there is an ongoing phase transformation as in BSCF at temperatures below 840 °C (cf. [4], Chapter 8).

Furthermore, D^{δ} and k^{δ} can only be extracted if conductivity equilibration does not occur faster than the achievable pO_2 step change. For instance, estimated equilibration times τ_1 for the LSCF sample at 800 °C calculated according to [11] using D^{δ} and k^{δ} values in air from literature [12] amount to ~15 s (if purely surface-exchange controlled behavior is assumed) or ~100 s (solely diffusion-controlled), which is similar to the pumping times of the setup in this pO_2 region: at 800 °C typical t_{90} values – after which 90 % of the pO_2 change is achieved – range from 40 seconds (performing a jump from 10^{-1} bar to 10^{-2} bar) to less than 10 seconds (for a jump from 10^{-4} bar to 10^{-5} bar).

6.3.4 k^{δ} Values Determined for Selected MIEC Compositions

Fig. 6.3 shows the k^{δ} values determined for LSCF at 800 °C in relation to literature data [12-14]. For the fitting, the final values of each reducing or oxidizing pO_2 jump were taken to plot the data. As can be seen, the values are in very good agreement to data reported for similar LSCF compositions by other research groups which consider LSCF in the given composition as a standard SOFC cathode material with very good oxygen-transport properties.

In Fig. 6.3, the values for the surface exchange coefficient of LSCF at 800 °C derived from the reducing pO_2 steps are shown in comparison with values derived from the oxidizing steps. It is apparent that there are slight differences between the surface exchange coefficients determined for oxidation and reduction runs, respectively. These deviations appear logical as the abrupt changes in oxygen concentration (by one order of magnitude) at the beginning of a pO_2 step are liable to strongly affect the oxygen surface exchange reaction. Given the different experimental conditions during (for instance) a reductive jump from an initial pO_2 of 10^{-2} bar to a final value of 10^{-3} bar, as compared with an oxidative jump from an initial pO_2 of 10^{-4} bar to the same final value of 10^{-3} bar (where, at the beginning of the jump, only one-hundredth of the gaseous oxygen available in the former reductive jump is present in the gas phase), the temporal redistribution of point defects within the solid is prone to take on a slightly different course, resulting in a different kinetic behavior, depending on the nature of the rate-determining steps for oxygen exchange.

Although a smaller step size would lessen the impact of this effect, this argument holds for most conductivity relaxation techniques. By applying near-equilibrium techniques such as electrochemical impedance spectroscopy (e.g., [5;14;15]) or an electrical conductivity relaxation (ECR) technique in the frequency domain (e.g., [16;17], Chapter 7), this difficulty can be overcome by staying close to chemical equilibrium and thus yielding only one single value for each pO_2. However, these methods are experimentally much more demanding and/or require well-adapted sample geometries in order to achieve similar results as $f(pO_2,T)$.

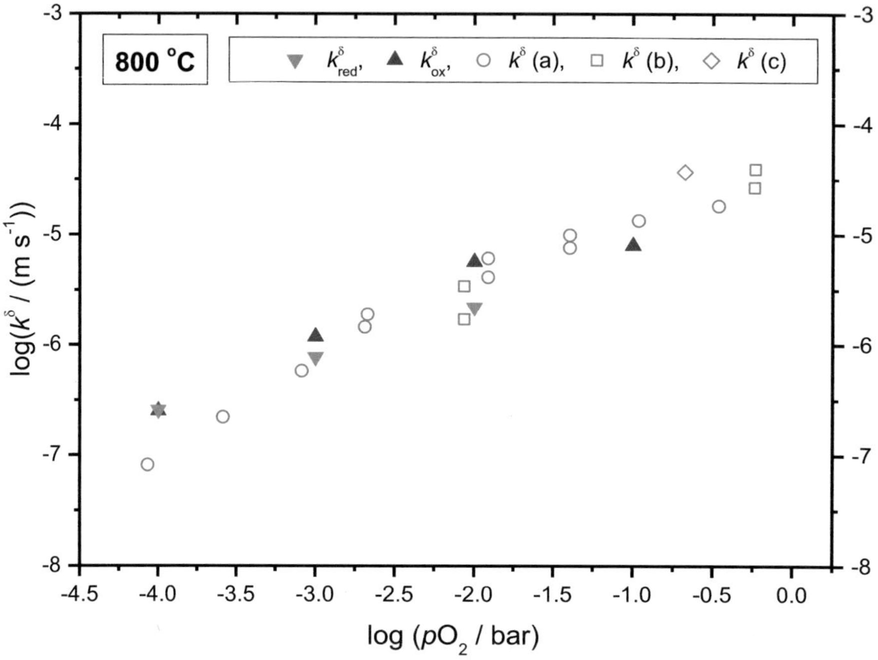

Fig. 6.3: k^δ values determined for LSCF at 800 °C in relation to literature data, (a) [12], (b) [13], and (c) [14]. Very good agreement is achieved for the values derived from oxidation and reduction runs of the measurement. Differences between values from oxidation runs to those from reduction runs are owed to the different kinetic behavior of the LSCF system at different oxygen partial pressures.

Under very oxidizing conditions ($pO_2 > 0.1$ bar) the process of oxygen pumping into or out of the sample chamber may take several minutes due to the vast amount of oxygen that must be transported through the YSZ tube (cf. Section 6.3.3). Compared to the oxygen exchange between sample and ambient gas phase, this is a very slow process. Therefore, the conductivity response follows the partial pressure change instantaneously, and neither D^δ nor k^δ values can be extracted (cf. Fig. 6.4).

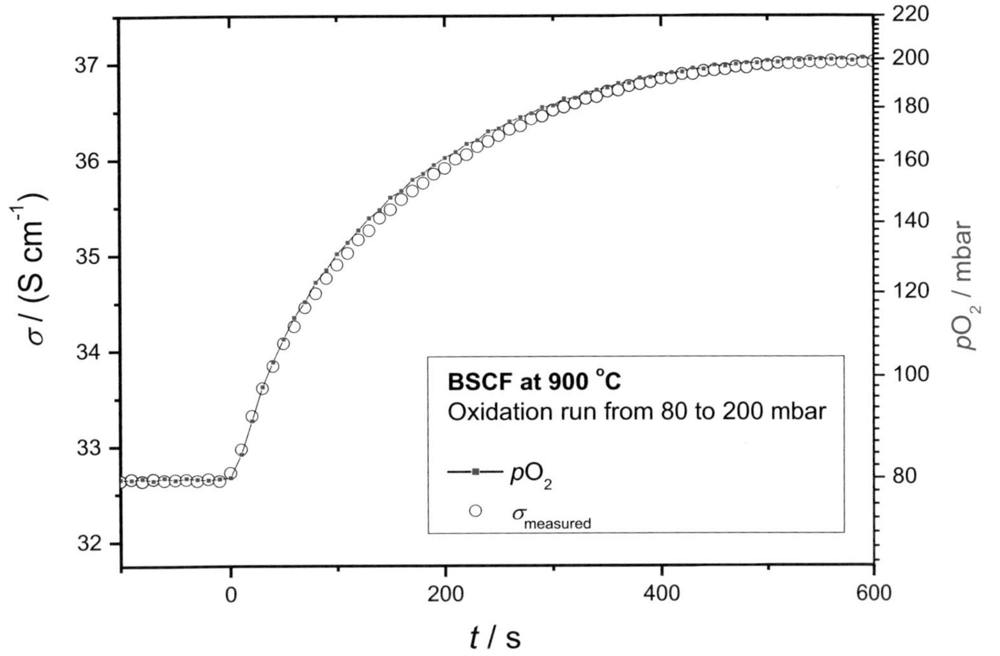

Fig. 6.4: Oxygen partial pressure (pO_2) change (blue symbols) for an oxidation run from 80 to 200 mbar at 900 °C in comparison to the conductivity (open symbols) determined for a BSCF sample (FZJ-BSCF-02). Sample equilibration kinetics is so fast that the conductivity signal exactly follows the pO_2 change achievable in the sample chamber under these highly oxidizing conditions. Therefore, neither D^{δ} nor k^{δ} values can be extracted.

The "oxygen pump" setup used here (cf. Chapters 2, 3) hence not only facilitates coulometric titration measurements on powder samples, thus providing knowledge on the stability limits of MIEC compositions as $f(pO_2)$ down to 10^{-18} bar, but, as shown here, allows for electrical conductivity measurements which in principle facilitate determination of D^{δ} and k^{δ} values within a certain pO_2 range. The lower limit of this range (at around 10^{-5} bar) is determined by the specific material properties – D^{δ} and k^{δ} values decrease, thus leading to long response times. The upper limit (at around 10^{-1} bar) is given by the speed limitations of the setup (oxygen pumping through the zirconia). In-between these two limiting pO_2 values, however, the determination of k^{δ} values is easily possible.

Fig. 6.5 shows k^{δ} values for all three MIEC compositions (BSCF, LSCF and PSCF) at 900 °C in the pO_2 range from 10^{-5} to 10^{-1} bar derived from the reducing steps. Due to the sluggish behavior of the compositions at partial pressures below approx. 10^{-4} bar, the reducing steps provided more data points in comparison to the oxidizing part of the measurement. Missing are data points for reducing steps where the fit process did not yield reliable results.

The data point represented by the blue star indicates the k^δ value for a porous LSCF solid oxide fuel cell (SOFC) cathode extracted from electrical impedance spectroscopy measurements as reported in our paper [5]. This value fits very well to the data extracted by ECR in this chapter, considering the concave run of $k^\delta(pO_2)$ for LSCF as reported in [12]. For BSCF, LSCF, and PSCF no pO_2-dependent kinetic parameters at this high temperature are available from literature so far.

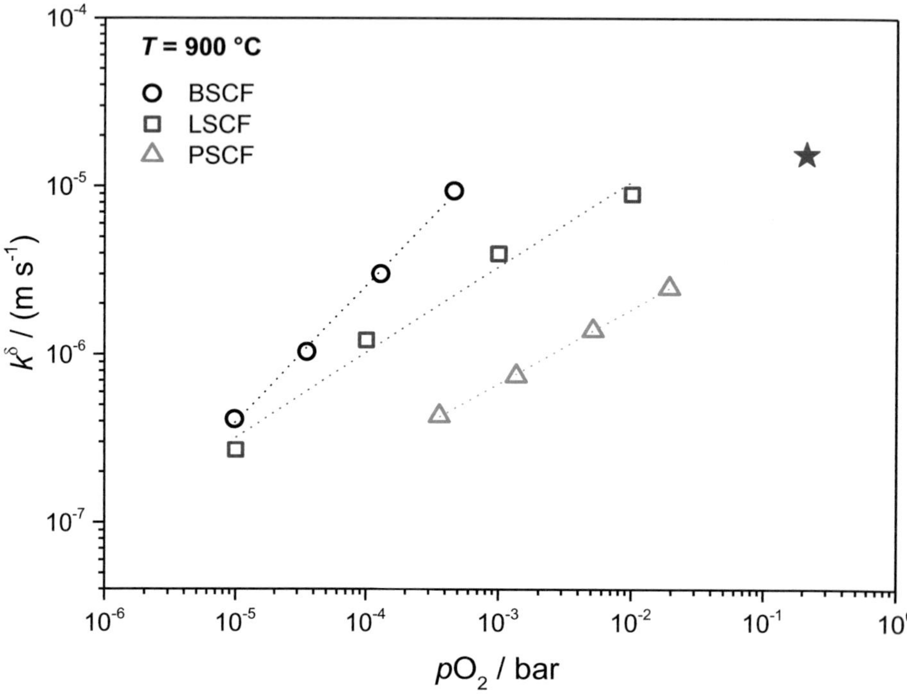

Fig. 6.5: Chemical oxygen surface exchange coefficients determined from measurements on ceramic bulk samples of BSCF, LSCF, and PSCF as a function of pO_2 at 900 °C from reductive pO_2 steps. The blue star indicates the value of $1.55 \cdot 10^{-5}$ m s^{-1} for LSCF at ambient pressure at 900 °C derived from electrical impedance spectroscopy on anode supported solid oxide fuel cells as reported in our paper [5].

At the lower temperature of 800 °C no D^δ or k^δ values were obtained in the case of BSCF due to its phase instability below 840 °C [4] and the resulting influence on conductivity; the PSCF sample – due to its apparent low D^δ and k^δ values even at 900 °C – behaved too sluggish at these lower temperatures, while the measurements on LSCF as shown in Fig. 6.3 demonstrated the viability of this measurement method by its very good agreement with literature data.

As expected, BSCF performs best, with k^δ values as high as 10^{-5} m s^{-1} (at an oxygen partial pressure of approx. 10^{-3} bar), followed by LSCF and PSCF in decreasing order. There is an indication that under highly oxidizing conditions the k^δ values of BSCF are much higher than those of LSCF, corresponding well with the published lower oxygen flow rates through LSCF OTMs that, however, also seem to be due to lower kinetic parameter values of LSCF under the reducing conditions on the low-pO_2 permeate side.

The strong decrease of the surface exchange parameter values at low oxygen partial pressures found for all examined MIEC materials leads to the conclusion that for thin high performance OTMs, the permeate side oxygen excorporation forms the bottleneck of the oxygen transport process. This has also been simulated by A. Häffelin for an LSCF membrane in our paper [6]. As we show there, further improvement in OTM performance can be achieved by greatly enhancing the surface area on the permeate side of the membrane by applying a nanoporous functional layer of the same or a different MIEC material. For BSCF, the conductivity measurements at even lower oxygen partial pressures discussed in Chapter 5 with the discovered equilibration times of several hundred hours suggest that transport parameters decrease even further with decreasing pO_2.

In the shown measurements, it was regrettably impossible to extract reliable values for the bulk diffusion coefficient D^δ for any of the examined material compositions. To visualize the reason for this, Fig. 6.6 shows the sum of error squares for all measurement points in comparison to calculated values derived from varying the parameter values around the fit results in a range of ±3 decades for both D^δ and k^δ.

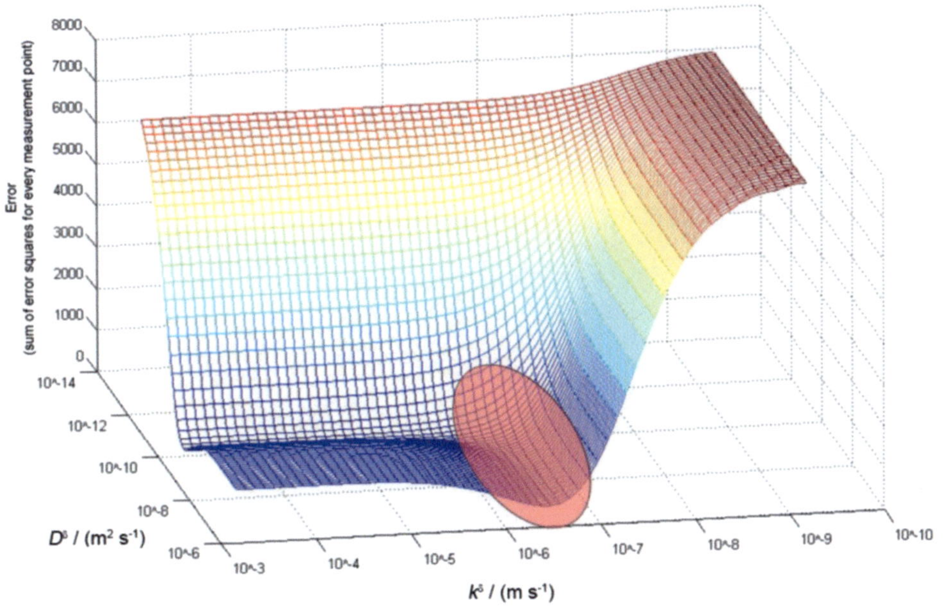

Fig. 6.6: Visualization of the error squares over all data points in comparison with mathematical calculations for a ±3 decade range in the D^δ and k^δ values around the initial fit result. The curved trench indicates possible combinations of ±3 decades for both D^δ and k^δ also resulting in relatively low error sums.

The initial fit result consisting of one pair of values for D^δ and k^δ possesses the smallest error and is represented by the 'deepest' ditch in the shown diagram surface, situated in the exact middle of the base k^δ-D^δ-plane. Easily visible is the fact that there exist other possible combinations of both parameters also resulting in relatively low errors, visualized as the trench in the plotted surface, one leg of which is marked with the red ellipse in Fig. 6.6.

To get a better visual impression of this predicament, the graph shown above has been reworked into a 2-dimensional k^δ-D^δ-plot, were for every combination of D^δ and k^δ the color of the corresponding data point is indicative of the amount of error resulting from the fit. The initially received fit values are again represented by the data point exactly in the middle of the k^δ-D^δ-plane. The error sum for this parameter tuple is then used to calculate ranges of 3, 10, 30 and 100 times this amount of error and every possible parameter variation's resulting error is colored accordingly in Fig. 6.7.

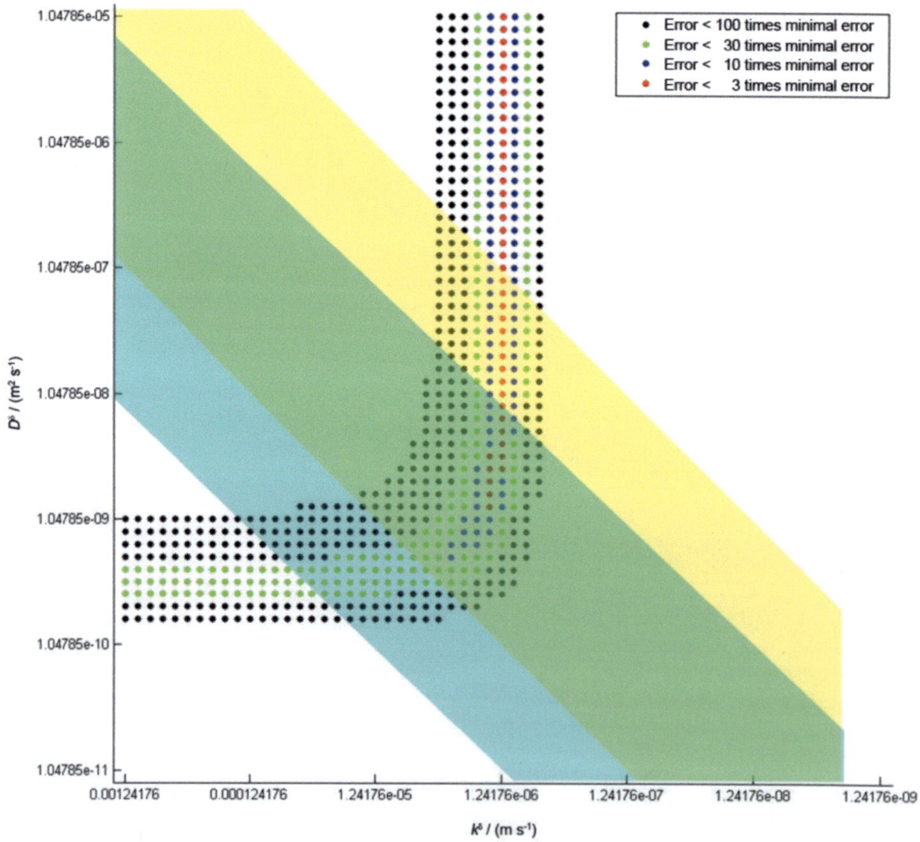

Fig. 6.7: Visualization of the error squares over all data points in comparison with mathematical calculations for a ±3 decade rage in the D^δ and k^δ values around the initial fit result. The colors indicate the individually calculated error sums in comparison to the error sum of the initial fit result.

Parameter tuples resulting in less than three times the lowest achievable error sum are displayed in red. It is easily visible that the value for k^δ can be determined with great accuracy, as a relatively small deviation from the optimal value immediately increases the error sum by a large amount, indicated by the blue, green and black colors. The value for the bulk diffusion coefficient D^δ, however, can be varied by several orders of magnitude while never resulting in a significant increase in error – all values higher than 10^{-9} m^2s^{-1} still yield results that stay in the range of three times the lowest possible error sum. In good conscience, this has to result in not extracting any value for D^δ. At the most, a lower limit of the parameter's expected value might be extracted.

This is an interesting find, as the criteria for the extraction of stable parameter values given by den Otter et al. in [12] are satisfied for the best fit result – both criteria are satisfied within the green colored area in Fig. 6.7 – so for this measurement, both D^δ and k^δ values should be obtainable. In this case, however, D^δ cannot be determined with accurate precision. It was found that this problem exists for all measurements, allowing only for the extraction of surface exchange values for all three MIEC materials at 900 °C.

In order to be able to also measure values for the bulk diffusion coefficient, it is necessary to adjust the sample dimensions – i.e. increase the thickness of the samples according to the newly acquired knowledge about the lower end of the possible value range. As thicker samples increase the amount of oxygen emitted during a reducing step in the oxygen pump, the pump parameters have to be adjusted to individually match the samples expected properties. Length of the sample can also be reduced in these cases, to decrease sample volume. In this case however, precautions have to be taken in order not to introduce too large an error into the conductivity measurement. Finally, smaller step sizes can be used to counteract the fact of a larger amount of released oxygen due to bigger sample volume – again necessitating individual PID-parameter adjustment. This work is currently ongoing.

6.4 Conclusions

By using the oxygen pump setup introduced in Chapters 2 and 3, reasonably fast pO_2 changes enable an application of the electrical conductivity relaxation (ECR) technique on MIEC oxides at temperatures above 700 °C, thus theoretically yielding D^δ and k^δ values as $f(pO_2, T)$.

- For BSCF, LSCF and PSCF bulk samples k^δ values between 10^{-5} and 0.1 bar pO_2 could be determined.
- The k^δ values determined for LSCF at 800 °C are in very good agreement with literature data, demonstrating the feasibility of this method.

- k^{δ} values for BSCF, LSCF, and PSCF could be determined for the first time for the high temperature of 900 °C.
- Owing to the pO_2 dependence of the transport parameters the conductivity changes of all examined MIEC samples become too sluggish at pO_2 values lower than 10^{-5} bar. This necessitates a change in sample geometry, e.g. measurements on dense thin films, in order to gain values under these reducing conditions where bulk samples take far too long to equilibrate.
- A visualization of the error sums for different parameter tuples enables a more refined determination of the validity of the fit results than previously suggested in literature.
- Determination of D^{δ} was not possible due to the samples being mainly surface controlled at the measuring conditions. Ongoing work looking into adaptation of sample size, step change magnitude and in every case specific required controller parameter adaptation will result in trustworthy bulk diffusion values.

References

[1] C. Niedrig, W. Menesklou, S. F. Wagner and E. Ivers-Tiffée, "High-Temperature pO_2 Stability of Metal Oxides Determined by Amperometric Oxygen Titration", *Journal of The Electrochemical Society* **160**, pp. F135-F140 (2013).

[2] C. Niedrig, L. Unger, S. F. Wagner, W. Menesklou and E. Ivers-Tiffée, "Electrical Conductivity Dependence on Oxygen Partial Pressure in Selected MIEC Materials for Oxygen Transport Membranes", manuscript in preparation (2015).

[3] M. Burriel, C. Niedrig, W. Menesklou, S. F. Wagner, J. Santiso and E. Ivers-Tiffée, "BSCF epitaxial thin films: Electrical transport and oxygen surface exchange", *Solid State Ionics* **181**, pp. 602-608 (2010).

[4] C. Niedrig, S. Taufall, M. Burriel, W. Menesklou, S. F. Wagner, S. Baumann and E. Ivers-Tiffée, "Thermal stability of the cubic phase in $Ba_{0.5}Sr_{0.5}Co_{0.8}Fe_{0.2}O_{3-\delta}$ (BSCF)", *Solid State Ionics* **197**, pp. 25-31 (2011).

[5] C. Endler-Schuck, J. Joos, C. Niedrig, A. Weber and E. Ivers-Tiffée, "The chemical oxygen surface exchange and bulk diffusion coefficient determined by impedance spectroscopy of porous $La_{0.58}Sr_{0.4}Co_{0.2}Fe_{0.8}O_{3-\delta}$ (LSCF) cathodes", *Solid State Ionics* **269**, pp. 67-79 (2015).

[6] A. Häffelin, C. Niedrig, S. F. Wagner, S. Baumann, W. A. Meulenberg and E. Ivers-Tiffée, "Three-Dimensional Performance Model for Oxygen Transport Membranes", *Journal of The Electrochemical Society* **161**, pp. F1409-F1415 (2014).

[7] J. Crank, *The mathematics of diffusion*, 2. ed., Oxford: Clarendon Press (1975).

[8] S. Švarcová, K. Wiik, J. Tolchard, H. J. M. Bouwmeester and T. Grande, "Structural instability of cubic perovskite $Ba_xSr_{1-x}Co_{1-y}Fe_yO_{3-\delta}$", *Solid State Ionics* **178**, pp. 1787-1791 (2008).

[9] D. N. Müller, R. A. De Souza, T. E. Weirich, D. Roehrens, J. Mayer and M. Martin, "A kinetic study of the decomposition of the cubic perovskite-type oxide $Ba_xSr_{1-x}Co_{0.8}Fe_{0.2}O_{3-\delta}$ (BSCF) (x=0.1 and 0.5)", *Physical Chemistry Chemical Physics* **12**, pp. 10320-10328 (2010).

[10] M. W. den Otter, *A study of oxygen transport in mixed conducting oxides using isotopic exchange and conductivity relaxation*, Ph.D. Thesis, Enschede: University of Twente (2000).

[11] M. W. den Otter, H. J. M. Bouwmeester, B. A. Boukamp and H. Verweij, "Reactor Flush Time Correction in Relaxation Experiments", *Journal of The Electrochemical Society* **148**, pp. J1-J6 (2001).

[12] H. J. M. Bouwmeester, M. W. den Otter and B. A. Boukamp, "Oxygen transport in $La_{0.6}Sr_{0.4}Co_{1-y}Fe_yO_{3-\delta}$", *Journal of Solid State Electrochemistry* **8**, pp. 599-605 (2004).

[13] P. Ried, E. Bucher, W. Preis, W. Sitte and P. Holtappels, "Characterisation of $La_{0.6}Sr_{0.4}Co_{0.2}Fe_{0.8}O_{3-\delta}$ and $Ba_{0.5}Sr_{0.5}Co_{0.8}Fe_{0.2}O_{3-\delta}$ as Cathode Materials for the Application in Intermediate Temperature Fuel Cells", *ECS Transactions* **7**, pp. 1217-1224 (2007).

[14] A. Leonide, *SOFC Modelling and Parameter Identification by means of Impedance Spectroscopy*, Karlsruhe: KIT Scientific Publishing (2010).

[15] A. Leonide, B. Rüger, A. Weber, W. A. Meulenberg and E. Ivers-Tiffée, "Impedance Study of Alternative $(La,Sr)FeO_{3-\delta}$ and $(La,Sr)(Co,Fe)O_{3-\delta}$ MIEC Cathode Compositions", *Journal of The Electrochemical Society* **157**, pp. B234-B239 (2010).

[16] C. Tragut and K. H. Härdtl, "Kinetic Behaviour of Resistive Oxygen Sensors", *Sensors and Actuators B-Chemical* **4**, pp. 425-429 (1991).

[17] S. F. Wagner, W. Menesklou, T. Schneider and E. Ivers-Tiffée, "Kinetics of oxygen incorporation into $SrTiO_3$ investigated by frequency-domain analysis", *Journal of Electroceramics* **13**, pp. 645-651 (2004).

Chapter 7

Kinetic Measurements on Thin Films

7.1 Introduction

In the previous chapter, the kinetics of oxygen transport have been studied in the oxygen pump setup on dense polycrystalline MIEC samples using electrical conductivity relaxation (ECR) in the time domain. While surface exchange parameters k^{δ} for the BSCF and LSCF samples could be extracted at partial pressures below 10^{-2} bar, the pump setup is too slow to accurately determine values around atmospheric pressure.

However, ECR measurements can also be performed in the frequency domain, where the fast kinetic behavior of samples is investigated in the pressure modulation setup first presented by Tragut et al. [4] and described in Chapter 2. For surface-controlled samples, the surface exchange parameter k^{δ} can then be determined according to the model described in more detail in [5].

Given the frequency range of the pressure modulations in the setup, fast samples – namely NGO supported thin films as already shown and discussed in chapters 4 and 5 – can be analyzed at intermediate temperatures between 400 °C and 700 °C. For these thin films, the oxygen permeation rate will purely be controlled by the surface exchange coefficient k^{δ} of the material. The use of dense epitaxial BSCF thin films will in theory permit extraction of the k^{δ} values undisturbed by grain boundary influences present in bulk samples. However, the strain and stress on the thin film due to the differences in cell parameters between BSCF and NGO may influence oxygen stoichiometry and conductivity behavior (cf. Chapters 4 and 5). Therefore it is of interest to compare the measurements with data extracted for the polycrystalline thin films also already discussed in the previous chapters.

Parts of the results shown in this chapter have been published in [1-3].

Furthermore, using this method on BSCF thin films, changes in the material's surface performance due to the presence of carbon dioxide will be easily visible.

7.2 Experimental

For the conductivity measurements two parallel gold contacts were painted at the surface of the epitaxial thin film samples (preparation cf. Chapter 2) and fired for one hour at a temperature of 900 °C in air. The MOD films were contacted with two parallel gold contacts by magnetron sputtering and subsequent spot-welding of Au-wires. Owing to the fairly high electrical resistance of the thin films, their planar resistance response was measured in two-point-technique.

ECR measurements in the frequency domain were performed in the kinetics measurement setup as described in Chapter 2 from 400 °C to 700 °C. Dwelling times after a change in temperature could be rather short – typically in the range of minutes – as sample equilibrium with the ambient air pressure is reached nearly instantaneously due to the very high surface exchange coefficient k^{δ} at the high oxygen partial pressure of 0.21 bar as well as the very small sample volume of the thin film material.

7.3 Results and Discussion

7.3.1 Surface Exchange Values for BSCF Thin Films

Fig. 7.1 shows the Bode plot of the frequency response function (normalized amplitude of the electrical conductance as a result of a pO_2 modulation) of a 340 nm thick epitaxial BSCF thin film on NGO in compressed synthetic air (CO_2-free) at six different temperatures from 450 °C to 700 °C (left) as well as a comparison of the data to the calculated transfer function for the measurement at 650 °C (right).

The descending slope of $\frac{d|G|}{df} \approx -1$ for all measurements as well as the very good correspondence to the calculated transfer function both show the purely surface controlled behavior of the sample as well as the viability of surface exchange parameter extraction.

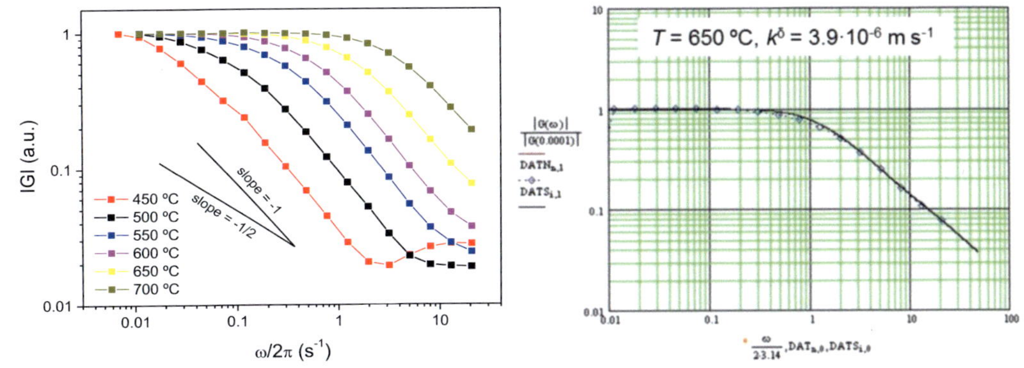

Fig. 7.1: Measured transfer functions for conductivity relaxation measurements on an epitaxial BSCF thin film (PLD-BSCF-17n1) at temperatures from 450 °C to 700 °C (left) and in comparison to the calculated transfer function for 650 °C (right).

In Fig. 7.2, the Arrhenius plot of the surface exchange coefficients k^δ extracted by fitting the frequency response function to the solution of Fick's law in the frequency domain with the given boundary conditions for surface-controlled behavior is shown for epitaxial BSCF thin films and polycrystalline MOD-derived thin films in comparison to measurements on bulk samples from literature.

The data for the epitaxial thin films was extracted from measurements on two 340 nm thick films on NGO. In order to achieve statistically sound values, every sample was measured from two to four times during a thermal cycle of first heating up in steps of 50 K from 450 °C to 700 °C and subsequently cooling down in identical steps, with short dwell times of 5 minutes after each temperature change for sample oxygen equilibration. The linear fit to all data points yields an activation energy of the surface exchange process of 0.96 eV.

For a 50 nm thick polycrystalline MOD thin film the same cycled measurement was performed four times from 400 °C up to 600 °C (colored symbols in Fig. 7.2). For this sample the transfer function also exhibited the same slope of approx. -1 as shown for the epitaxial sample in Fig. 7.1. The activation energy of the surface exchange process is calculated to be 1.35 eV.

A comparison to literature values for the activation energies measured by ECR in the time domain for polycrystalline bulk samples between 550 °C and 700 °C is inconclusive. Girdauskaite et al. [6] found an activation energy of 1.66 eV, whereas Bucher et al. determined a value of 0.66 eV [7]. This rather large discrepancy can on one hand originate from the various difficulties with the ECR technique described in Chapter 6 (i.e. thermal sample history, different surface treatment etc.), on the other hand the extraction of accurate surface exchange values for bulk samples at these – comparatively low – temperatures is very difficult. Combined with measurement temperatures up to 750 °C, where phase changes are occurring in the material (cf. Chapter 8), this could explain said discrepancy as well as the comparatively extensive scattering in the data from Bucher et al. [7].

Between epitaxial and polycrystalline films measured similarly, however, a comparison is viable. The higher activation energy is exhibited by the polycrystalline BSCF film, suggesting that the influence of the substrate is, in this case, lessened in comparison to the epitaxial film and oxygen exchange can occur more easily, resulting in higher k^{δ} values with increasing temperature. This is in accordance with the findings on the different conductivity behavior with temperature as described in Chapter 4.

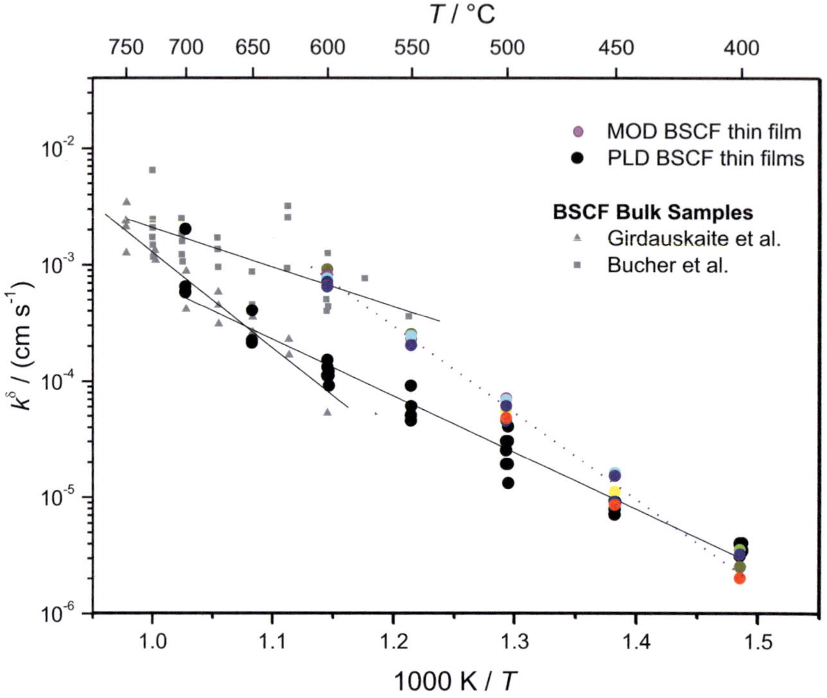

Fig. 7.2: Measured values for the surface exchange coefficient k^{δ} of a MOD BSCF thin film (MOD-BSCF-068, colored symbols) and epitaxial BSCF thin films (PLD-BSCF-17n1 and PLD-BSCF-18n1, black symbols) at different temperatures in comparison to literature data for bulk samples from Girdauskaite et al. [6] and Bucher et al. [7] (grey symbols).

While for epitaxial thin films – as can be seen in Fig. 7.1 (left) – the slope in the Bode plot is very close to the ideal slope of -1 that could be expected for purely surface controlled kinetic behavior (cf. Chapter 2), the polycrystalline MOD-derived thin films behave slightly different, exhibiting an oxygen exchange behavior that results in a marginally less steep decline of the transfer function (black squares in Figs. 7.3 and 7.4). As discussed in a previous paper [5] and mentioned in Chapter 2, the analysis of electrical conductivity relaxation data in the frequency domain conveniently facilitates distinction between surface or diffusion controlled kinetics (slopes of -1 or -½ in the Bode plot, respectively) for the case of an ideal homogeneous sample morphology (cuboid single crystal). In the case of a polycrystalline thin film, completely ideal behavior is not to be expected. Therefore, due to polycrystallinity and (possibly) surface morphology, this model represents a first-order approximation. Hence, even predominant surface control may not exhibit the ideal -1 slope.

Nonetheless, the model still yields valid results as is discussed in the following: The influence of bulk diffusion (D^{δ}) can be neglected as is shown in Fig. 7.3. For the fixed k^{δ} value of $1.4 \cdot 10^{-5}$ cm/s a variation of D^{δ} by several decades around the expected value of $\sim 10^{-7}$ cm²/s (obtained by extrapolating data from [6] down to lower temperatures) only start to show any influence on the fit function at values of well below 10^{-8} cm²/s (only at lower values diffusion control would become observable, visible by a significant change in the slope, ultimately converging towards a value of -1/2).

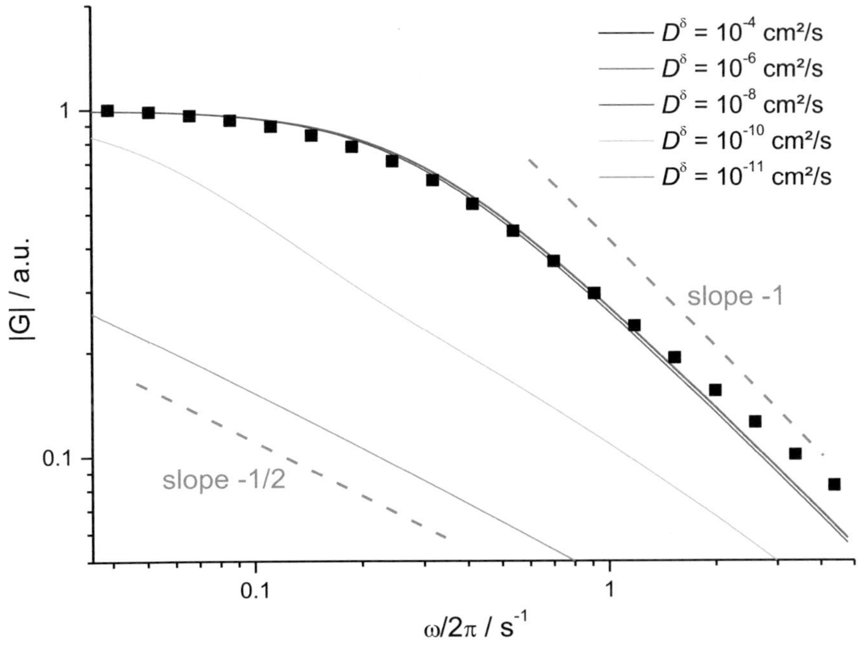

Fig. 7.3: Measured (normalized) transfer function for sample MOD-BSCF-068 at 450 °C in comparison to mathematical fits for $k^{\delta} = 1.4 \cdot 10^{-5}$ cm/s and variations of several orders of magnitude around the extrapolated bulk diffusion coefficient D^{δ}.

On the other hand, as is shown in Fig. 7.4, even slight changes in the k^{δ} value (by a factor of 2) can immediately be discerned in the given case of surface control.

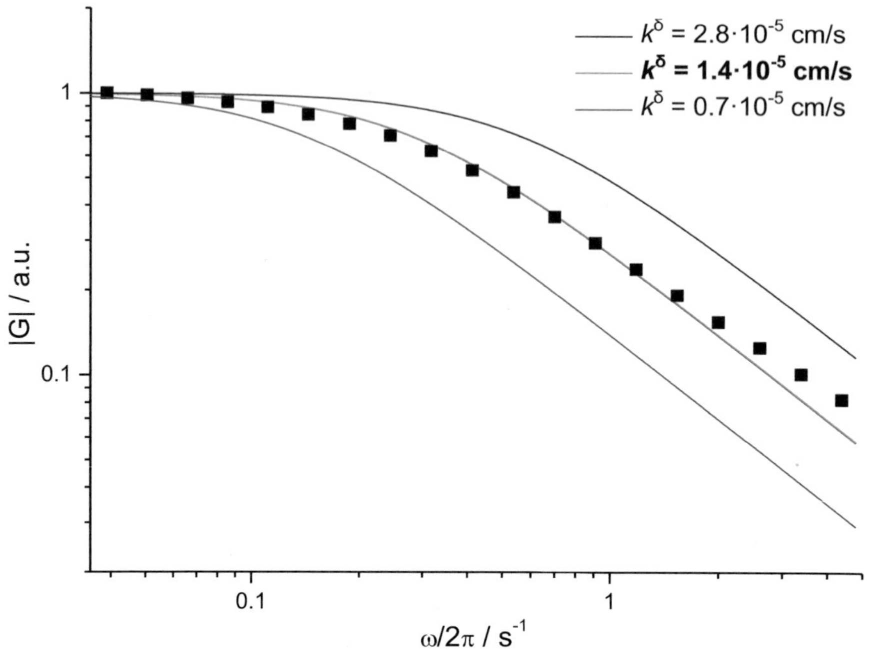

Fig. 7.4: Measured (normalized) transfer function for sample MOD-BSCF-068 at 450 °C in comparison to mathematical fits for $D^\delta = 10^{-7}$ cm²/s and only small variations around the extracted value for the surface exchange coefficient k^δ.

7.3.2 Surface Exchange Values for LSCF Thin Films

Similar measurements have been conducted on LSCF thin films on NGO. Fig. 7.5 shows a comparison of the surface exchange values for 1 µm thick sputtered thin films and an MOD derived thin film (thickness 125 nm) with data for bulk samples (derived from ECR measurements in the time domain) and an SOFC cathode (derived by electrical impedance spectroscopy, EIS) from literature.

For the 1 µm thick sputtered films, no reliable values could be extracted above 600 °C, as the samples response became too fast for the 30 Hz limitation of the measurement setup. Higher temperatures could be measured for the MOD derived film, as its absolute surface exchange values were approx. one order of magnitude lower by comparison. This is due to the sample morphology of the differently produced thin films (see SEM micrographs below). The activation energies for all three thin film samples, however, are pretty similar between the thin films themselves (at 1.01 eV for the MOD thin film and 0.88 eV as well as 1.05 eV for both sputtered thin films) and in comparison to literature values extracted from impedance spectroscopy measurements on porous LSCF cathodes by A. Leonide (1.05 eV) [8]. This suggests that the influence of the NGO substrate on the polycrystalline films is negligible for the LSCF/NGO system, as stoichiometry changes in the LSCF material are much less pronounced than for BSCF (cf. Chapter 5).

The very good agreement of the values measured for the MOD thin film with literature data in Fig. 7.5 suggests that the comparably high values derived for the sputtered samples stem from different sample properties.

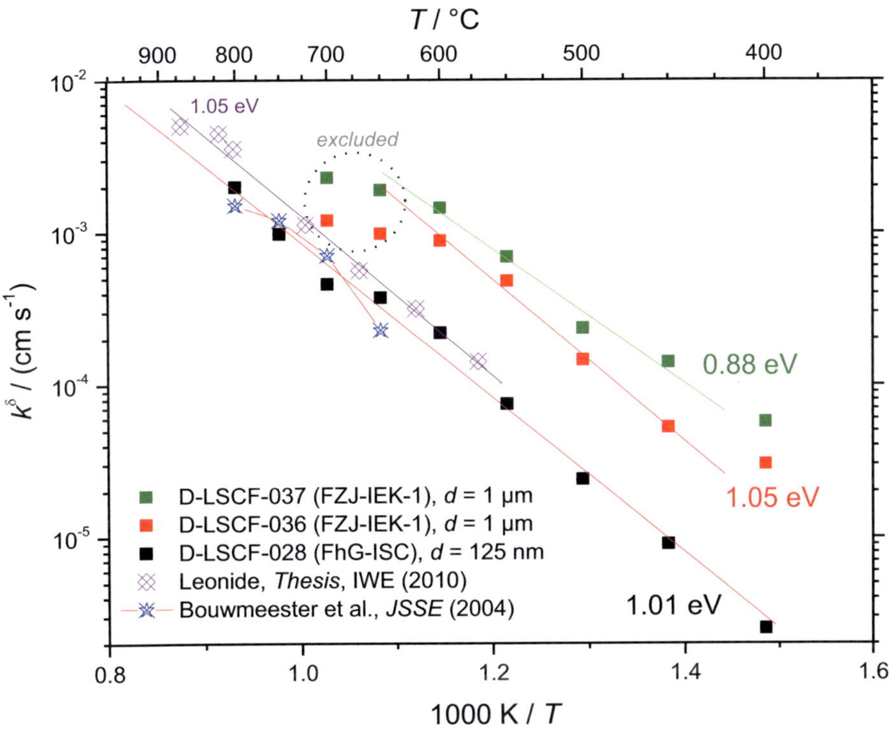

Fig. 7.5: Measured values for the surface exchange coefficient k^{δ} of sputtered LSCF thin films (D-LSCF-037 and D-LSCF-036, green and red squares) and a MOD LSCF thin film (D-LSCF-028, black squares) at different temperatures in comparison to literature data (colored hollow symbols) from Leonide [8] and Bouwmeester et al. [9]. The four data points labeled "excluded" have been omitted for the pseudo-linear fit due to the fast sample response reaching the kinetic limitations of the measurement setup.

Fig. 7.6 shows a comparison between the surfaces of the MOD derived thin film (left) and one of the nearly identical sputtered samples (right). While the MOD process yielded a somewhat smooth, dense surface, the sputtering produced a layer consisting of single LSCF grains, accounting for an enlargement of the active surface area. The latter sample should therefore produce larger values for the effective surface exchange coefficient. However, this fact alone cannot account for the differences of nearly one order of magnitude between the measured k^{δ} values for both systems.

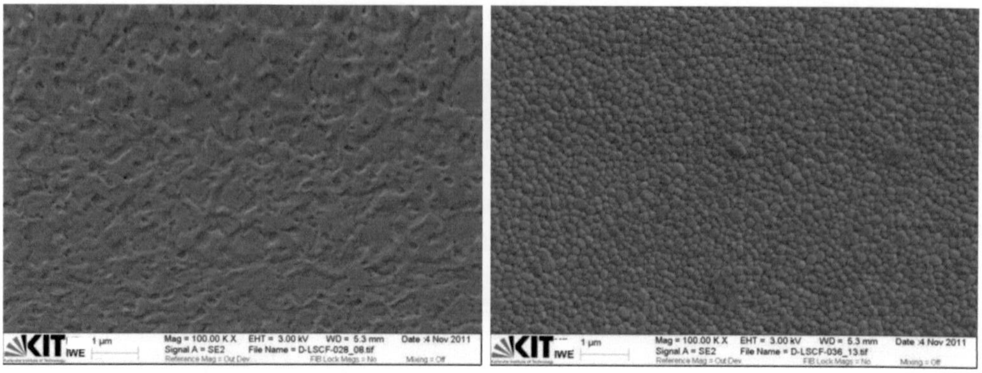

Fig. 7.6: Surface morphology depicted by SEM for the MOD LSCF thin film (left) and a sputtered thin film (right). The very grainy surface of the latter accounts for a larger active surface area.

While a post-test cross-section of the MOD sample showed the LSCF layer to be completely dense, the sputtered layer turned out to be porous in nature, with a high probability for open porosity (Fig. 7.7).

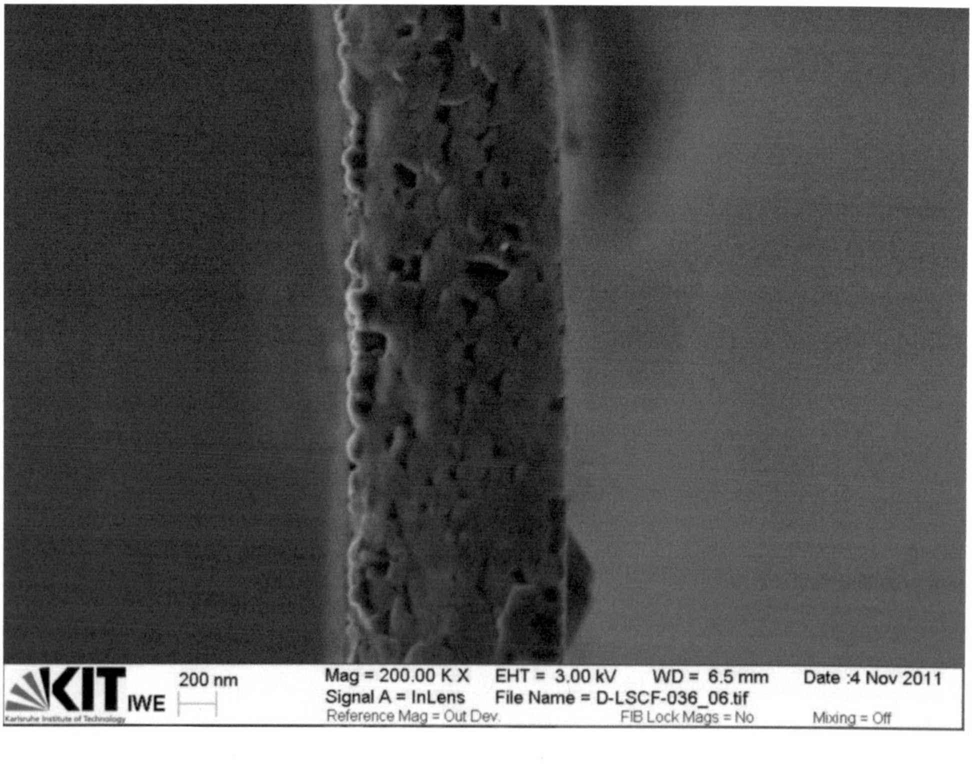

Fig. 7.7: Cross section of one sputtered LSCF layer on NGO. The substrate is to the left of the image. Porosity is clearly visible. The slight curving of the layer is an artifact due to sample charging by the scanning electron beam.

Surface morphology together with porosity can easily account for the difference in calculated values for k^δ and the lower values derived from the dense MOD thin film should more accurately represent the true LSCF performance as also indicated by the good agreement with literature data for higher temperatures.

7.3.3 Degradation of the BSCF Surface by Elevated CO_2 Levels

While high CO_2 contents of the flue gas in a possible 4-end module installation may not be detrimental to the performance of an LSCF membrane [10], BSCF is prone to form carbonates at high concentrations of carbon dioxide [10-13], and even at very low concentrations of 0.04 % CO_2, Bucher et al. report carbonate formation at the surface [14]. With the shown ECR technique, measurements on thin films should readily detect changes in the membrane surface by degrading surface exchange coefficients.

To this end, a measurement series was performed on an epitaxial BSCF thin film between 400 °C and 600 °C in three different gas mixtures, namely synthetic air containing no CO_2 at all (as in all the measurements shown above), the same gas mixture containing 0.1 % CO_2 and a third one containing 1 % CO_2.

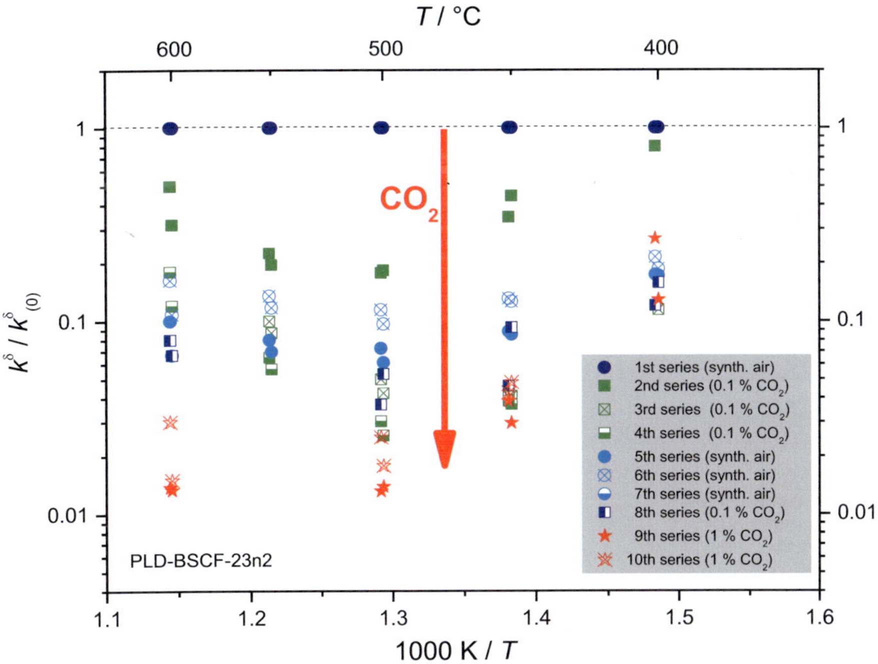

Fig. 7.8: Degradation of the surface exchange coefficient k^δ normalized to the initial (CO_2 free) surface exchange coefficient $k^\delta_{(0)}$ in synthetic air containing 0 %, 0.1 % and 1 % CO_2 at temperatures between 400 °C and 600 °C.

The measurement was performed in 50 K steps with first increasing and then decreasing temperature for each measurement series. The first series was a ‘baseline’ measurement in CO_2-free synthetic air, followed by three consecutive series in 0.1 % CO_2, then three series in pure synthetic air again (to test for eventual reversibility), then one more series in 0.1 % CO_2, followed by two measurement series in 1 % CO_2 containing synthetic air.

Fig. 7.8 shows the degradation of the extracted surface exchange coefficient k^δ relative to the very first measurement series’ results $k^\delta_{(0)}$. The clearly visible initial degradation during the 2nd measurement series (0.1 % CO_2) intensifies in the following two series (3rd and 4th) while the latter two nearly yield the same performance with only a slight difference in the extracted k^δ values (green symbols). During the next three series (5th to 7th) in CO_2 free atmospheres, a small increase is visible in the surface exchange coefficient, most pronounced at the highest measured temperatures, probably due to decreasing carbonate content at the sample surface. Returning to 0.1 % CO_2 in the measurement gas for the 8th measurement series does not reduce the performance significantly more than already seen in the third exposure to that gas composition in series 4. During the next two series in 1 % CO_2, however, at temperatures above 450 °C, an even more pronounced decrease in surface exchange rates is found (red symbols), amounting to a total decrease of approx. two orders of magnitude in comparison to the initial values.

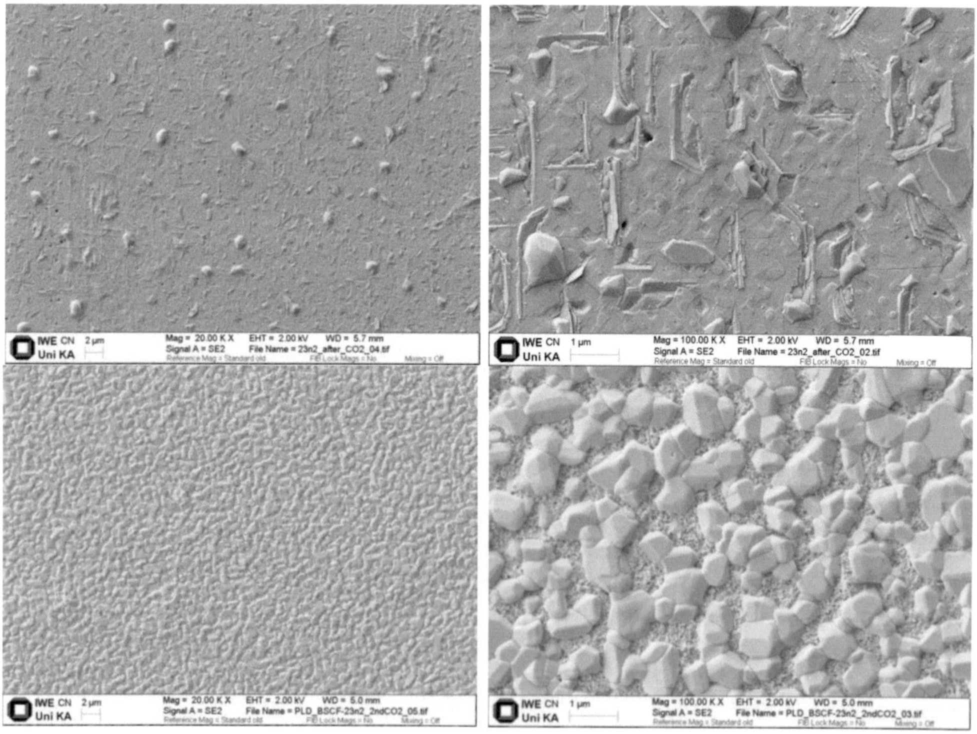

Fig. 7.9: Changes in surface morphology of the formerly epitaxial BSCF thin film (PLD-BSCF-23n2) due to repeated measurements in synthetic air containing firstly 0.1 % CO_2 (top row), followed by 1 % CO_2 (bottom row).

Post-test analyses of the sample surface by SEM and EDX are shown in Figs. 7.9 and 7.10. Clearly visible are the morphological changes like coarsening and polycrystalline grain growth in the formerly epitaxial thin film due to the CO_2 treatment.

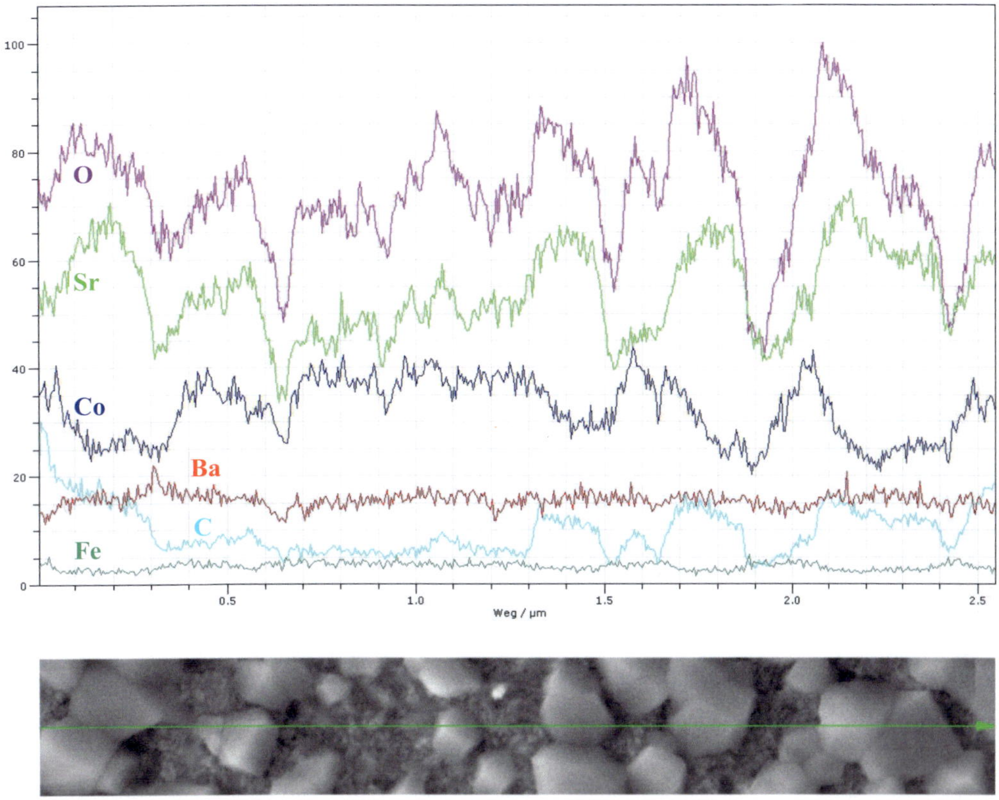

Fig. 7.10: Energy dispersive X-ray analysis in the SEM: EDX linescan over a surface area of the sample after the 10 measurement series shown in Fig. 7.8. Element content analysis (top) clearly shows the presence of carbon (light blue curve) in the visible large grains. The path over the sample surface is shown in the bottom image.

The element specific analysis by EDX (Fig. 7.10) shows the formation of polycrystalline carbonate grains after measurements in atmospheres with a low CO_2 content. This has been verified by a subsequent transmission electron microscopic (TEM) analysis of the sample shown in Fig. 7.11, performed by S. Roitsch from the Ernst-Ruska-Centre (ER-C) at the Forschungszentrum Jülich, Jülich, Germany. Even low concentrations of CO_2 – thrice and thirty times the amount present in ambient air – are seemingly sufficient to severely damage the surface of the BSCF thin film by carbonate formation.

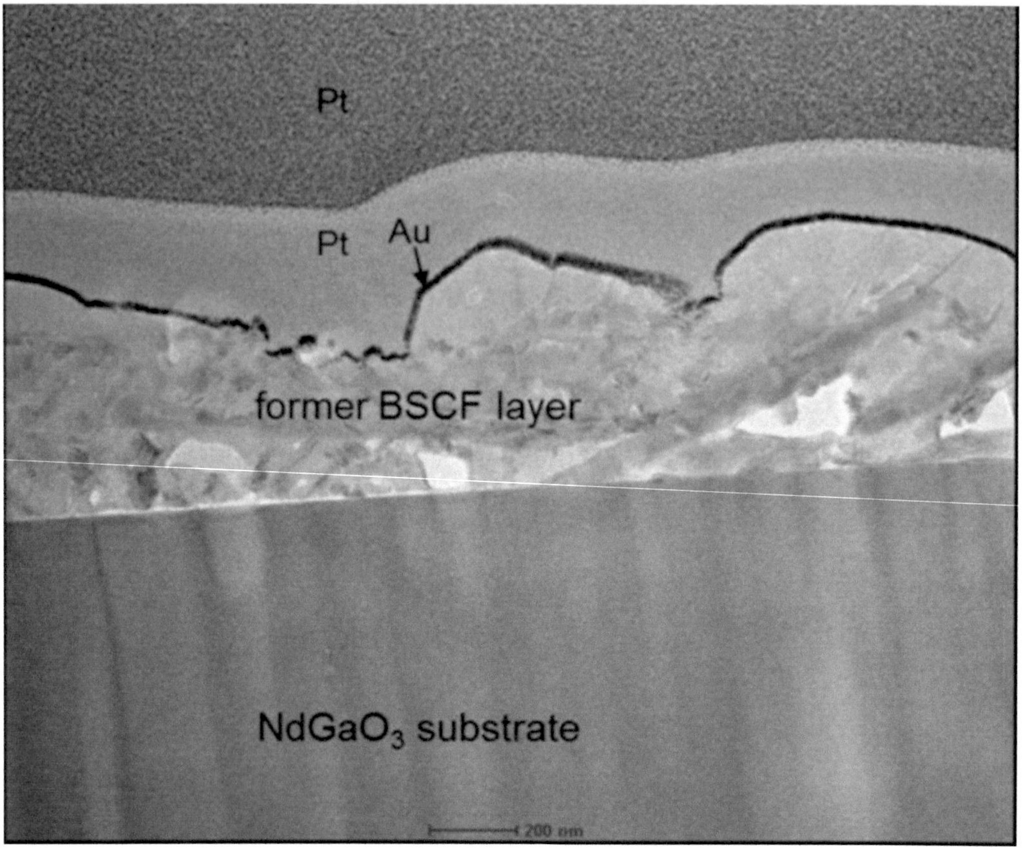

Fig. 7.11: TEM micrograph of a FIB lamella prepared from the sample after the 10 measurement series shown in Fig. 7.8. The epitaxial BSCF layer has been completely destroyed (picture taken by S. Roitsch, ER-C, Forschungszentrum Jülich, Jülich, Germany).

In order to further verify this degradation in atmospheres with a CO_2 content of only 1 %, BSCF powder has been annealed at 800 °C for six days in a constant flow of pure oxygen containing this small amount of CO_2. The longer treatment and higher temperature – compared to the measurement conditions for the thin film sample – are owed to the comparably large grain size (D_{50} = 2.40 µm) of the powder, where small surface reactions only become visible in the XRD analysis if they account for a certain fraction of the sample volume. A comparison of the results to untreated BSCF powder is shown in Fig. 7.12.

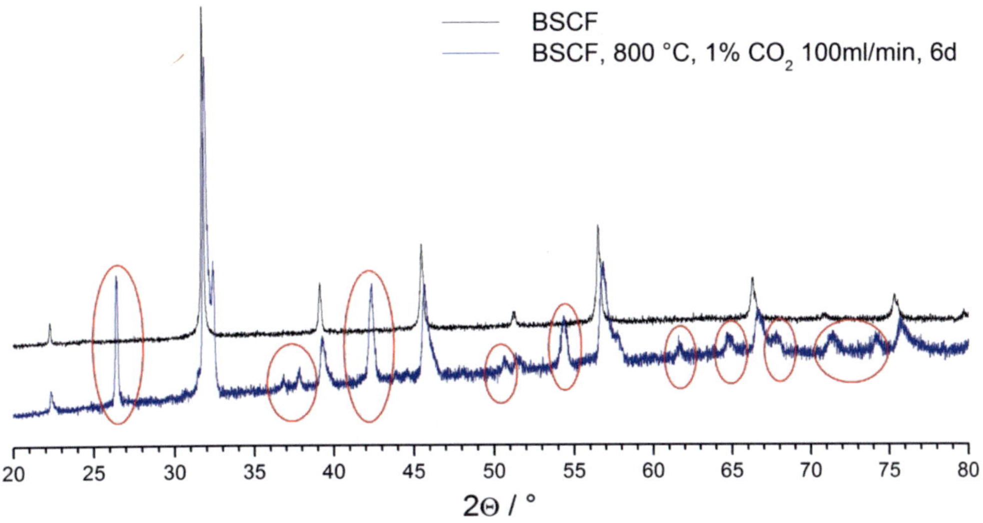

Fig. 7.12: XRD comparison of purely cubic BSCF powder (black curve) with BSCF containing secondary phases after annealing in 1% CO_2 (blue curve). No background subtraction has been performed in order to avoid unintended smoothing of low intensity secondary peaks. Newly formed reflections – representing hexagonal phase (cf. Chapter 8) and decomposition products – are marked in red.

While the three most pronounced secondary phase peaks (at 26.5°, 42.5° and 54.5°) in the treated powder represent the formation of the hexagonal phase in BSCF (as discussed in Chapter 8), all other newly formed reflections are due to decomposition of BSCF because of ongoing reactions with carbon dioxide.

7.4 Conclusions

In this chapter PLD-derived epitaxial BSCF thin films and polycrystalline thin films from both BSCF (MOD) and LSCF (MOD and sputtering) with their intrinsic properties as purely surface-controlled MIEC systems have been investigated by ECR in the frequency domain using the kinetic setup described in Chapter 2 and the surface exchange parameters k^δ have been determined in synthetic air between 400…700 °C.

- The model formerly developed for single-crystalline MIEC systems [15] is applicable for parameter extraction on these samples, hence enabling near-equilibrium surface exchange parameter extraction as opposed to (by design) non-equilibrium measurements described in Chapter 6.

- Despite the mathematical model being developed with the premise of measuring single-crystalline samples – which is expectedly ideally met by epitaxial thin films as indicated by the perfect slope of -1 in the Bode plot – applicability could also be shown for the polycrystalline samples exhibiting only a slight deviation from the ideal slope.
- For epitaxial BSCF thin films the activation energy for the surface exchange process calculated to 0.96 eV.
- Polycrystalline MOD BSCF thin films exhibited an activation energy of the surface exchange process of 1.35 eV.
- The higher activation energy exhibited by the polycrystalline film gives further evidence to the assumption (previously formulated in Chapter 4) of lessened substrate influence – and therefore better oxygen exchange at higher temperatures – in comparison to the epitaxial system.
- For polycrystalline MOD LSCF thin films the activation energy for the surface exchange process calculated to 1.01 eV, very similar to 1.05 eV found for LSCF SOFC cathodes [8]. This suggests negligible substrate influence for the LSCF/NGO system.
- Sputtered LSCF thin films yielded activation energies of 0.88 and 1.05 eV while the resulting k^{δ} values proved to be unexpectedly high (one order of magnitude higher than all other data reported), suggesting possible open porosity present in the samples, as also indicated by an SEM analysis.

Furthermore, by using ECR in the frequency domain in this kinetic setup, changes in the thin film's surface performance due to the presence of selectable gaseous contaminants can be monitored in situ. This was done for BSCF epitaxial thin films concerning low concentrations of carbon dioxide in synthetic air:

- BSCF thin films degrade in atmospheres containing even small amounts of CO_2 (thrice the concentration of ambient air). Carbonate formation proved to be partly reversible, however, at higher amounts (already at 1 % CO_2) the thin film was irreversibly destroyed.

References

[1] M. Burriel, C. Niedrig, W. Menesklou, S. F. Wagner, J. Santiso and E. Ivers-Tiffée, "BSCF epitaxial thin films: Electrical transport and oxygen surface exchange", *Solid State Ionics* **181**, pp. 602-608 (2010).

[2] C. Endler-Schuck, J. Joos, C. Niedrig, A. Weber and E. Ivers-Tiffée, "The chemical oxygen surface exchange and bulk diffusion coefficient determined by impedance spectroscopy of porous $La_{0.58}Sr_{0.4}Co_{0.2}Fe_{0.8}O_{3-\delta}$ (LSCF) cathodes", *Solid State Ionics* **269**, pp. 67-79 (2015).

[3] K. Asano, C. Niedrig, W. Menesklou, S. F. Wagner and E. Ivers-Tiffée, "Modification of Oxygen / $(Ba_{0.5}Sr_{0.5})(Co_{0.8}Fe_{0.2})O_{3-\delta}$ Interfaces Derived By Metal-Organic Deposition", *ECS Transactions* **66**, pp. 147-155 (2015).

[4] C. Tragut and K. H. Härdtl, "Kinetic Behaviour of Resistive Oxygen Sensors", *Sensors and Actuators B-Chemical* **4**, pp. 425-429 (1991).

[5] S. F. Wagner, W. Menesklou, T. Schneider and E. Ivers-Tiffée, "Kinetics of oxygen incorporation into $SrTiO_3$ investigated by frequency-domain analysis", *Journal of Electroceramics* **13**, pp. 645-651 (2004).

[6] E. Girdauskaite, H. Ullmann, V. V. Vashook, U. Guth, G. B. Caraman, E. Bucher and W. Sitte, "Oxygen transport properties of $Ba_{0.5}Sr_{0.5}Co_{0.8}Fe_{0.2}O_{3-x}$ and $Ca_{0.5}Sr_{0.5}Mn_{0.8}Fe_{0.2}O_{3-x}$ obtained from permeation and conductivity relaxation experiments", *Solid State Ionics* **179**, pp. 385-392 (2008).

[7] E. Bucher, A. Egger, P. Ried, W. Sitte and P. Holtappels, "Oxygen nonstoichiometry and exchange kinetics of $Ba_{0.5}Sr_{0.5}Co_{0.8}Fe_{0.2}O_{3-\delta}$", *Solid State Ionics* **179**, pp. 1032-1035 (2008).

[8] A. Leonide, *SOFC Modelling and Parameter Identification by means of Impedance Spectroscopy*, Karlsruhe: KIT Scientific Publishing (2010).

[9] H. J. M. Bouwmeester, M. W. den Otter and B. A. Boukamp, "Oxygen transport in $La_{0.6}Sr_{0.4}Co_{1-y}Fe_yO_{3-\delta}$", *Journal of Solid State Electrochemistry* **8**, pp. 599-605 (2004).

[10] S. A. Möbius, *Charakterisierung perowskitischer Hochtemperaturmembranen zur Sauerstoffbereitstellung für fossil gefeuerte Kraftwerksprozesse*, Dissertation, Fakultät für Maschinenwesen, RWTH Aachen (2010).

[11] M. Arnold, H. Wang and A. Feldhoff, "Influence of CO_2 on the oxygen permeation performance and the microstructure of perovskite-type $(Ba_{0.5}Sr_{0.5})(Co_{0.8}Fe_{0.2})O_{3-\delta}$ membranes", *Journal of Membrane Science* **293**, pp. 44-52 (2007).

[12] A. Waindich, A. Möbius and M. Müller, "Corrosion of $Ba_{1-x}Sr_xCo_{1-y}Fe_yO_{3-\delta}$ and $La_{0.3}Ba_{0.7}Co_{0.2}Fe_{0.8}O_{3-\delta}$ materials for oxygen separating membranes under Oxycoal conditions", *Journal of Membrane Science* **337**, pp. 182-187 (2009).

[13] A. Yan, M. Cheng, Y. L. Dong, W. S. Yang, V. Maragou, S. Q. Song and P. Tsiakaras, "Investigation of a $Ba_{0.5}Sr_{0.5}Co_{0.8}Fe_{0.2}O_{3-\delta}$ based cathode IT-SOFC I. The effect of CO_2 on the cell performance", *Applied Catalysis B-Environmental* **66**, pp. 64-71 (2006).

[14] E. Bucher, A. Egger, G. B. Caraman and W. Sitte, "Stability of the SOFC cathode material $(Ba,Sr)(Co,Fe)O_{3-\delta}$ in CO_2-containing atmospheres", *Journal of The Electrochemical Society* **155**, pp. B1218-B1224 (2008).

[15] C. Tragut, *Kinetik schneller Sauerstoffsensoren*, Düsseldorf: VDI Verlag GmbH (1992).

Chapter 8

Thermal Stability of BSCF

8.1 Introduction

In the past, there has been surprisingly little knowledge about the long-term structural and electrochemical stability of BSCF under application-relevant temperatures. Even though it is well-known that BSCF is not stable when operated in atmospheres containing carbon dioxide due to the formation of carbonates (cf. Chapter 7 and [6-9]) – a fact putting a question mark on the applicability of BSCF cathodes for SOFC operation in ambient air –, little has been reported on the structural stability of the cubic perovskite phase in BSCF, notably on the temperature-phase relationship, until recently.

A slow exponential decrease of the oxygen permeation rate with time (over ~ 400 h), caused by a phase transition, was already reported by Shao et al. [10] when BSCF membranes were operated at temperatures below 825...850 °C, and more recently, e.g., by Efimov et al. [11] at 750 °C. It has been observed that the cubic perovskite phase is not stable in air below these temperatures; a reversible phase change from a cubic into a two-phase mixture of hexagonal and cubic perovskites occurs [12]. According to Arnold et al. [13], this change would take place via a new monoclinic perovskite.

Indications for this phase change can for example be found in conductivity measurements as shown in Chapter 5. Fig. 8.1 shows a two-step increase in the electrical conductivity of a BSCF bulk sample after an oxidizing step in pO_2. While the fast response indicated by (1) is due to the oxygen uptake of the sample, a much slower second conductivity increase becomes visible (2). This can be attributed to a slow phase change in the BSCF lattice.

Parts of the results shown in this chapter have been published in [1-5].

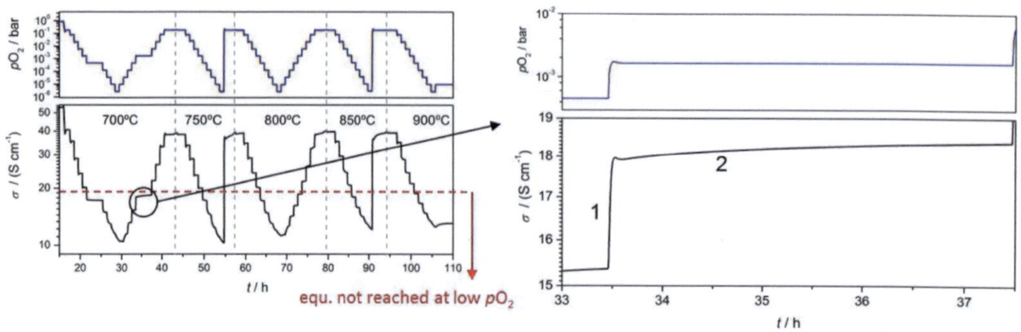

Fig. 8.1: Conductivity measurement on a BSCF bulk sample at different temperatures and oxygen partial pressures (left, from Fig. 5.1 in Chapter 5). During the longer dwell time at 10^{-3} bar oxygen partial pressure at 700 °C (circled area, enlarged on the right) a second, slow conductivity change becomes visible (2) after the initial sample response to the partial pressure change (1).

A much more visible representation of occurring structural changes is provided by scanning electron microscopy images of epitaxial BSCF thin films on NGO (cf. Chapters 2, 4, 5, 7) after treatment in different conditions concerning temperature and/or oxygen partial pressure. In Fig. 8.2, morphology changes become evident: while after the initial annealing treatment for 10 hours at 700 °C the thin film consists of an epitaxial layer riddled with plate-like structures (left), these distinctive structures completely vanish after burning the Pt contacts for 1 hour at 1050 °C (middle). The morphology changes further (right) during the measurements performed at different temperatures and pO_2 (cf. Chapter 5).

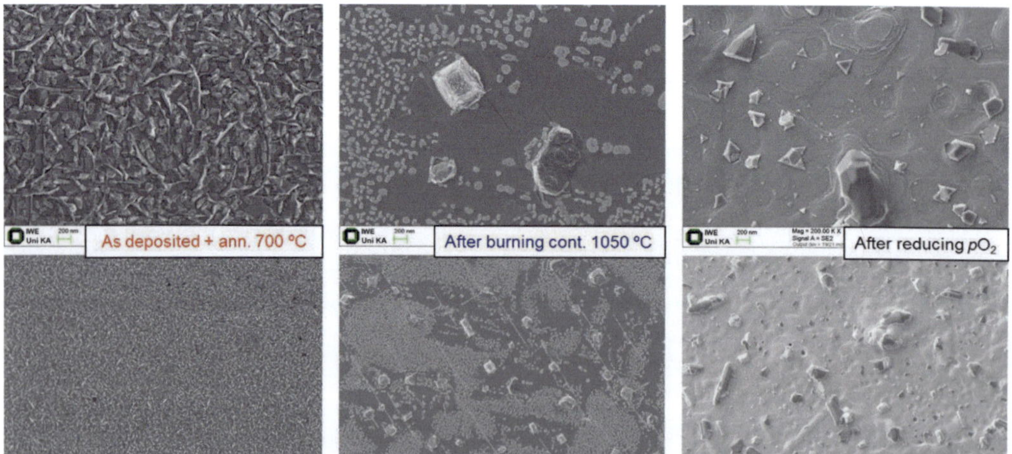

Fig. 8.2: Structural changes visible on an epitaxial thin film (PLD-BSCF-11a) after different temperature and/or oxygen partial pressure treatments (left: after annealing for 10 h at 700 °C, middle: after subsequent Pt-contacting and burning at 1050 °C, right: after conductivity measurements at different T and pO_2).

Fig. 8.3 provides a closer look at a similarly treated thin film sample. During the conductivity measurements at temperatures between 700 °C and 900 °C, the same plate-like structures have been formed.

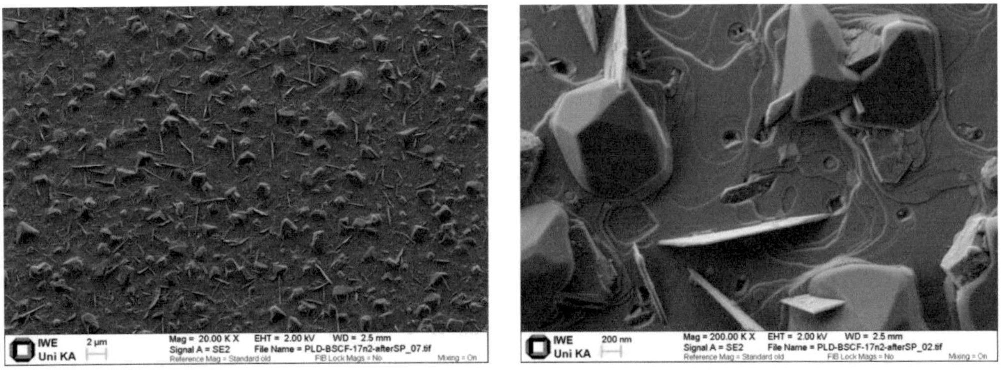

Fig. 8.3: SEM pictures of an epitaxial BSCF thin film after conductivity measurements at temperatures between 700 ° C ≤ T ≤ 900 °C and 10^{-6} bar ≤ pO_2 ≤ 1 bar (cf. Chapter 5).

In order to have a closer look at the formed structures, a pristine epitaxial thin film has been annealed for 10 hours at 900 °C in ambient air. Fig. 8.4 again shows the formation of plate-like secondary phases (left) reaching all the way down to the NGO substrate, visible in a FIB-lamella (right) prepared[2] from the section marked in red.

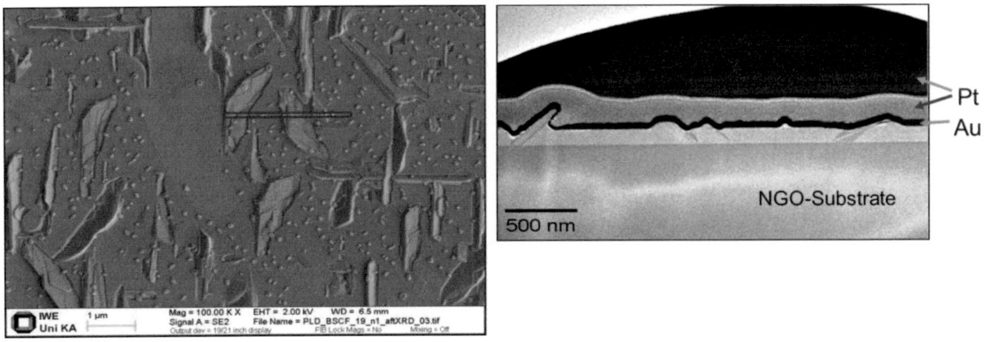

Fig. 8.4: SEM picture of an epitaxial BSCF thin film (PLD-BSCF-19n1) annealed for 10 hours at 900 °C in ambient air (left) and FIB lamella prepared from the section marked in red (right). FIB preparation and TEM image recorded by S. Roitsch, ER-C, Forschungszentrum Jülich, Jülich, Germany.

Subsequent EDX analysis could shed some light on the chemical composition of the different visible structures (cf. Fig. 8.5). While the layer has been confirmed to be stoichiometrically correct BSCF5582, the plate-like structures appear strontium-deficient. Small precipitations at the

[2] Kindly provided by S. Roitsch, ER-C, Forschungszentrum Jülich, Jülich, Germany

surface contain less barium, probably compensating for the formation of the Ba-rich plate structures. A more detailed structural analysis of secondary phase formation in BSCF bulk samples will be shown in the second part of this Chapter.

element	atom%
O	58
Fe	5
Co	13
Sr	17
Ba	8

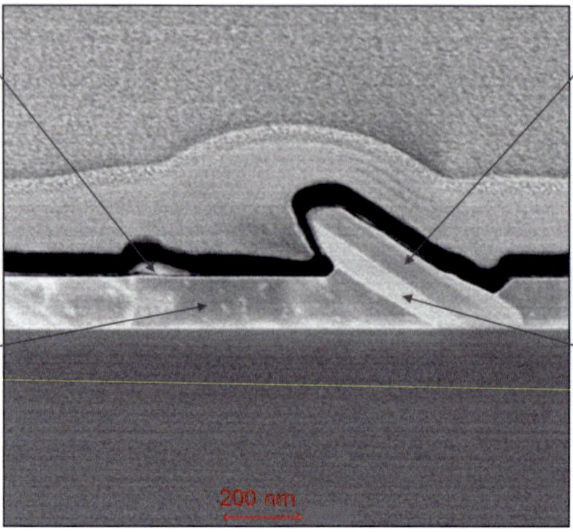

element	atom%
O	54
Fe	5
Co	27
Sr	4
Ba	11

element	atom%
O	56
Fe	5
Co	16
Sr	12
Ba	10

element	atom%
O	60
Fe	3
Co	18
Sr	4
Ba	14

Fig. 8.5: EDX analysis of different structures visible in the lamella from Fig. 8.4 (by S. Roitsch, ER-C, Forschungszentrum Jülich, Jülich, Germany).

Recently, there have been several further accounts in literature reporting on the phase composition of BSCF in the intermediate temperature range (around 800 °C). D. Müller et al. [14] were able to analyze the formation kinetics of the hexagonal BSCF phase at 800 °C by a combined XRD/electron microscopy study, resulting in a detected phase equilibrium between hexagonal and cubic perovskite phase at temperatures of 800 °C. By means of TEM analysis performed by P. Müller, we reported on a yet not unambiguously identified phase present in BSCF at intermediate temperatures, beside the expected cubic and hexagonal polymorphs [15]. Efimov et al. [16] very recently presented a detailed TEM study, identifying a decomposition of cubic BSCF into not only the hexagonal phase but also further phases in the intermediate temperature range below ~850 °C.

As the oxygen-transport properties, especially on a long-term time scale, are very sensitive to changes in material and/or phase composition, investigations concerning the phase stability of BSCF and the corresponding transport properties are highly anticipated, thus addressing issues of further performance optimization in the operating temperature range (700...900 °C).

Parts of the investigations shown in this chapter have been published recently [1-5].

8.2 Experimental

8.2.1 Powders

All BSCF powder experiments shown in this chapter were done using the $Ba_{0.5}Sr_{0.5}Co_{0.8}Fe_{0.2}O_{3-\delta}$ powder provided by the Fraunhofer Institute for Ceramic Technologies and Systems (IKTS), Hermsdorf/Germany (cf. Chapter 2). The D_{50} value of the powder was around 2.40 µm, XRD analysis (cf. Chapter 2 and graph (a) in Fig. 8.10) showed the powder to be of the single phase cubic perovskite structure with a lattice constant of 3.986 Å.

Annealing of the BSCF powders in ambient air was done using chamber furnaces, while treatment in synthetic air was performed in a tube furnace with a mixture of 20 % oxygen in nitrogen. Heating and cooling rates have been fixed to 20 K/min.

8.2.2 Electrically Contacted Bulk Samples with Different Grain Sizes

The bulk samples shown in this chapter were prepared from the single-phase cubic BSCF raw powders at the Institut für Energie- und Klimaforschung (IEK-1), Forschungszentrum Jülich, Jülich, Germany, by uniaxial pressing at approx. 10 kN/cm² and sintering for 12 hours at 1000 °C, for 50 h at 1120 °C and 50 h at 1150 °C (heating and cooling rates 5 K/min), resulting in mean grain sizes around 25 µm, 50 µm and 90 µm, respectively (Fig. 8.6, left to right), as determined by analyses using the software SPIP (Image Metrology A/S, Denmark). These bulks were subsequently cut by ultrasonic lapping and mechanically polished (P1200 emery paper), yielding bulk samples with a geometry of 10 x 5 x 1 mm³.

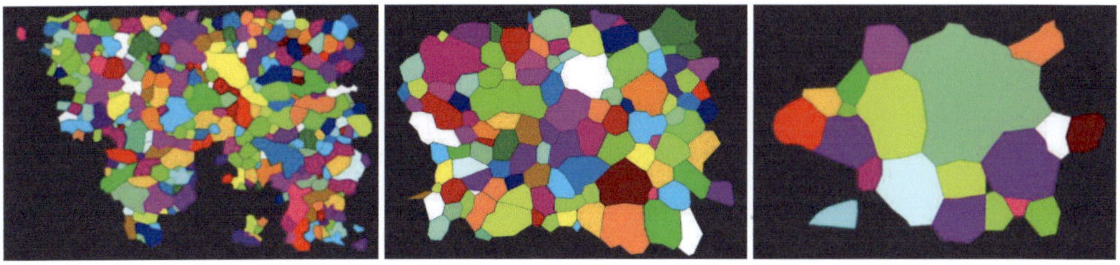

Fig. 8.6: SPIP analyses of bulks consisting of different grain sizes. Image width in each case represents 740 µm.

For conductivity measurements the bulks were contacted with Pt wires using a frit-free Pt paste fired at 1050 °C for 1 h in ambient air. Owing to the fairly low electrical resistance of the BSCF bulks, electrical measurements were carried out in 4-point technique (cf. Section 8.2.3).

8.2.3 Conductivity Measurements

Long-term conductivity measurements in ambient air have been conducted on the BSCF bulks in a chamber furnace. In this setup, five four-wire contacted samples can be measured at the same time by applying a serial current using a Keithley 224 programmable current source while sequentially reading the voltage over each sample's contacts using a Keithley 2000 multimeter. As described in Chapter 2, a possible sample degradation over time due to ionic demixing as a result of long-term DC loading could be ruled out by a careful comparison of long-term measurements on the same set of samples, one with a continuous current, the other with a setup that only applied a measurement current for one minute every hour to get a set of data points. Both experiments yielded the same rate of degradation.

8.2.4 SEM, TEM and EDX Analyses

All SEM, TEM and EDX analyses shown in Sections 8.4 and 8.5 were conducted by P. Müller and M. Meffert at the Laboratory for Electron Microscopy (LEM), Karlsruhe Institute of Technology (KIT), Karlsruhe, Germany and have been published in our joint paper [2].

BSCF bulk samples ($5 \cdot 5 \cdot 2\ mm^3$) were prepared from the single-phase cubic BSCF raw powders by isostatic pressing at 100 MPa, followed by a sintering step at 1000 °C for 12 h. In a second step the samples were homogenized at 1000 °C for 10 or 24 h to ensure that the samples contain only the cubic BSCF phase and subsequently annealed at the desired temperature for 100 h in ambient air and then quenched in water (Fig. 8.7).

Subsequently, the annealed samples were sent to P. Müller at the Laboratory for Electron Microscopy (LEM, KIT), cut in half and – as preparation for the detailed microstructure characterization by scanning electron microscopy (SEM) – polished and chemically etched using a colloidal silicon dioxide solution (Oxid Polier Suspension, Oberflächentechnologien Dieter Ziesmer, Germany). The SEM images at the LEM were taken with a Zeiss 1530 Gemini microscope using the in-lens detector. For transmission electron microscopy (TEM), the samples were prepared by conventional preparation methods involving grinding, dimpling, polishing and Ar^+-ion etching. TEM and STEM analyses were conducted using a Philips CM200FEG/ST and an FEI Titan 80-300 microscope operated at 200 and 300 keV respectively. Additionally, in [2], high-angle annular dark-field (HAADF) STEM images directly visualize different phase regions using the atomic-number contrast [17-19] of this imaging mode.

The EDX analyses at LEM, KIT were performed by P. Müller, using a 30 mm^2 EDAX Si(Li) detector system of the institute's FEI Titan 80-300 microscope with an energy resolution of 136 eV in order to determine the chemical composition of different phases. Composition quantification of the spectra was done with the FEI "TEM imaging and analysis" (TIA) software (Version: 4.3 build 904), as described in detail in our paper [2].

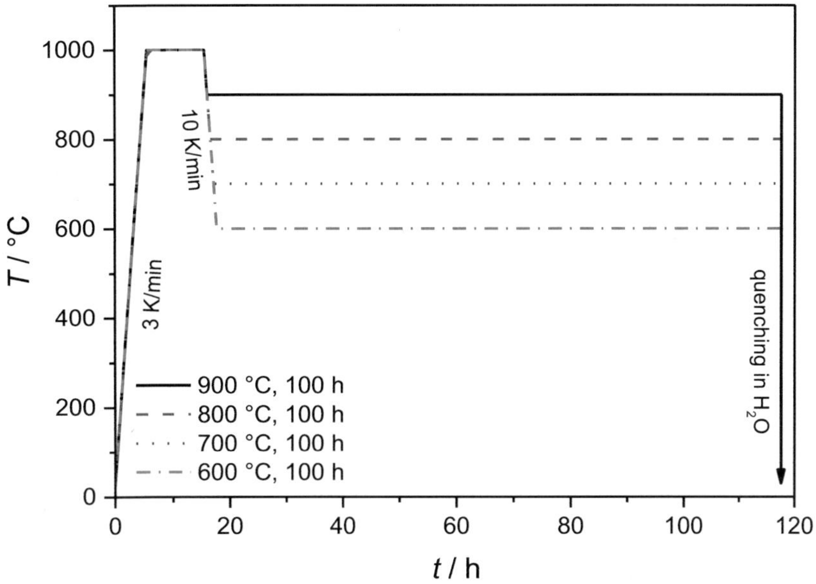

Fig. 8.7: Temperature treatment of BSCF bulk samples at IWE for subsequent SEM/TEM analyses at LEM.

8.3 Long Term Performance Degradation and Powder Analyses

8.3.1 Long Term Conductivity Measurements

The black curve in Fig. 8.8 shows the conductivity of a BSCF bulk sample measured in ambient air over several hundred hours each at temperatures varying from initial 900 °C, over 800 °C to finally 700 °C. The different behavior of the conductivity as a function of time at these three temperatures will be discussed below. First, however, the conductivity data have to be normalized.

Due to the highly changeable oxygen nonstoichiometry of the BSCF lattice at these high temperatures, resulting in substantial oxygen exchange with the ambient gas phase:

$$\frac{1}{2}O_{2,gas} + V_O^{\bullet\bullet} \rightleftarrows O_O^{x} + 2h^{\bullet} \qquad (9.1)$$

where O_O^x denotes oxygen ions on the regular oxygen lattice sites (in Kröger-Vink notation), the concentration of holes $h^\bullet$ and, thus, (p-type) electrical conductivity σ is changed with changing oxygen vacancy concentration. Therefore, variations in the oxygen partial pressure pO_2 as a result of ambient total air pressure variations during long-term measurements lead to corresponding variations in electrical conductivity.

Taking this into account, the slight fluctuations in the recorded conductivity data (Fig. 8.8) on a timescale of mere days can be related to meteorological changes in total ambient air pressure, displayed as the red curve. The corresponding behavior is clearly visible during steep meteorological changes, e.g. at values around approx. 1050 hours and 2200 hours into the measurement.

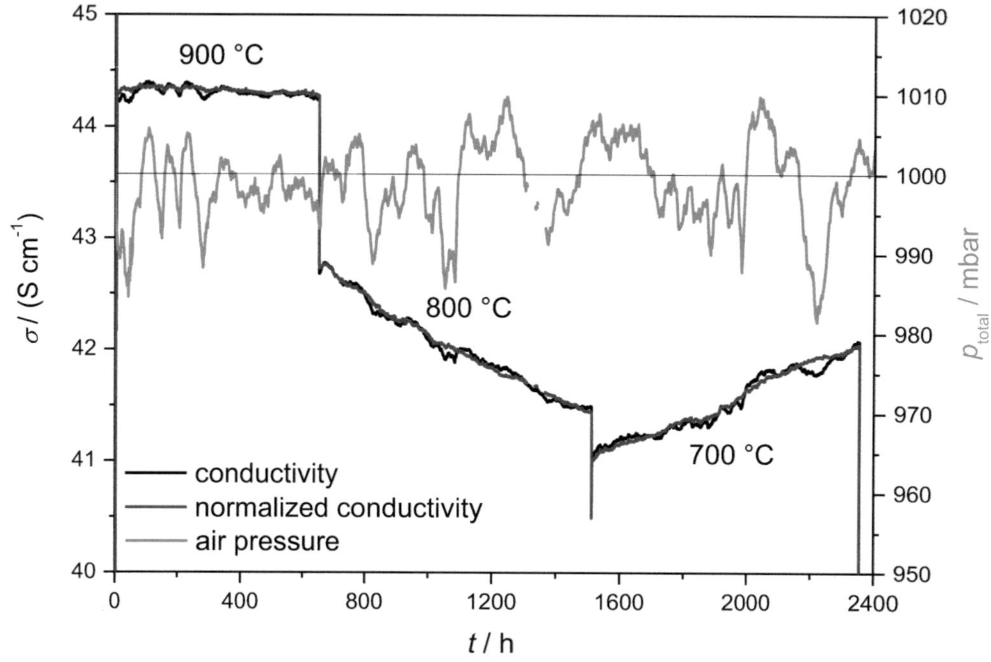

Fig. 8.8: Long term conductivity measurement of a BSCF bulk ceramic (FZJ-BSCF-05). At 800 °C and 700 °C, very pronounced changes in the conductivity can be observed. Original data (black) have been normalized to an ambient air pressure of 1000 mbar (blue) using meteorological data[3] (red curve).

In an attempt to normalize the conductivity to a certain 'constant' atmospheric pressure, the dependency of the electrical conductivity on oxygen partial pressure measured and discussed in Chapter 5 and our paper [20] can be used.

[3] Meteorological data were kindly provided by the Institute for Meteorology and Climate Research (IMK), KIT, Karlsruhe/Germany.

Fig. 8.9 shows a double logarithmic graph of the conductivity of BSCF at various oxygen partial pressures at 700 °C. As there are only small meteorological changes around 1000 mbar, these data can be fitted pseudo-linear in the range from 6 mbar < pO_2 < 200 mbar in order to calculate the conductivity at a certain air pressure using

$$\sigma_{th}(p) = 10^{\left(\lg(\sigma_0(1000\ \text{mbar})) + b \cdot \lg\left(\frac{p}{1000\ \text{mbar}}\right)\right)}, \quad (9.2)$$

where σ_{th} is the theoretical conductivity at a given total air pressure p. The theoretical difference $\Delta\sigma_{\text{th}}$ of the conductivity at a given p to the calculated conductivity at a fixed air pressure, e.g. 1000 mbar is

$$\Delta\sigma_{\text{th}}(p) = \sigma_{\text{th}}(p) - \sigma_0(1000\ \text{mbar}) \quad (9.3)$$

By scaling this difference to the measured conductivity of the long term experiments sample, the difference in the measurement to the normalized pressure of 1000 mbar calculates to

$$\Delta\sigma_{\text{real}} = \Delta\sigma_{\text{th}} \cdot \frac{\sigma_{\text{meas}}(p,t)}{\sigma_0(1000\ \text{mbar})}, \quad (9.4)$$

and therefore the corrected conductivity becomes

$$\sigma_{\text{real}} = \sigma_{\text{meas}}(p,t) - \Delta\sigma_{\text{real}}. \quad (9.5)$$

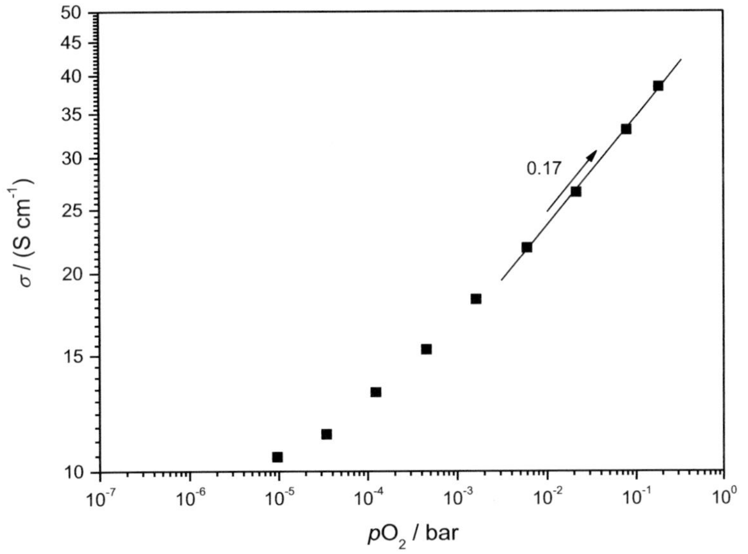

Fig. 8.9: Conductivity of BSCF at various oxygen partial pressures at $T = 700$ °C, fitted in the range from 6 mbar < pO_2 < 200 mbar (from measurements on FZJ-BSCF-02).

The thus obtained normalized conductivity is shown as the blue curve in Fig. 8.8. During the first 650 hours at 900 °C there is no decrease in conductivity (a change of -0.1 % is most likely attributed to the not quite perfect correction to ambient pressure changes). As electrical measurements on polycrystalline samples are very sensitive to phase composition and demixing due to cation diffusion, this indicates that the BSCF system remains as stable cubic perovskite at this temperature.

After cooling to 800 °C, a constant decrease of the electrical conductivity is clearly visible, amounting to 3.0 % of the initial value after holding the temperature for 860 hours, indicating occurring changes in the system, probably related to the formation of a hexagonal phase as reported e.g. by [12].

Interestingly, after further cooling to 700 °C, conductivity increases again, gaining 2.4 % compared to the new base value over 840 hours. If these changes in conductivity are indications for transition to the hexagonal phase, either the process occurring at 800 °C reverses at the lower temperature of 700 °C, or at least the phase equilibrium that will finally be present in the BSCF system at both temperatures is different.

8.3.2 Phase Changes in BSCF Powders

8.3.2.1 Growth of Hexagonal Phase at 800 °C

In order to analyze the occurring changes to the single phase cubic structure in the BSCF system, a batch of cubic BSCF powder was annealed in ambient air at 800 °C, rapidly (20 K/min) cooling down after periods of 72 h, 120 h, 240 h, 600 h and 1680 h, respectively, taking a sample for subsequent XRD analysis and heating up to 800 °C again. Fig. 8.10 shows the ongoing formation of the hexagonal phase from the initially purely cubic powder (a) to a mixture of both phases after 1680 hours (f), while all hexagonal reflections – corresponding to the (101), (201) and (112) planes – are already clearly detectable after 72 hours and continue to gain in intensity with amounting annealing time.

A WinINDEX fit (Bruker AXS) to the hexagonal peaks in the 2Θ-range from 20° to 80° yielded the hexagonal lattice constants to be $a,b = 5.629$ Å, $c = 4.361$ Å, in good agreement with the values determined by e.g. [12].

In an attempt to determine the kinetics of this occurring phase change, the intensities reflected by the two different lattice structures were used. Under the premises that a growing volume fraction of the hexagonal phase will coincide with an increase in reflected intensity represented by the corresponding peaks in the XRD spectra, an analysis of XRD peak area changes was undertaken.

The intensity of the radiation reflected by all the cubic planes decreased to 78.9 % of the initial value over the complete 1680 hours, while that obtained from only the main (110) reflection (calculated in a 2Θ-range from 31.2° to 32.5°) decreased to 79.6 %. As this is in good agreement considering the intrinsic accuracy of XRD measurements, for the analysis of all annealed powders only the main reflections – by far the most pronounced and therefore with the best signal-to-noise ratio – have been used, i.e. the cubic (110) reflection for analysis of cubic phase change and the hexagonal (101) reflection for analysis of hexagonal phase growth, respectively. The changes in reflected intensity by both planes are shown in Fig. 8.11.

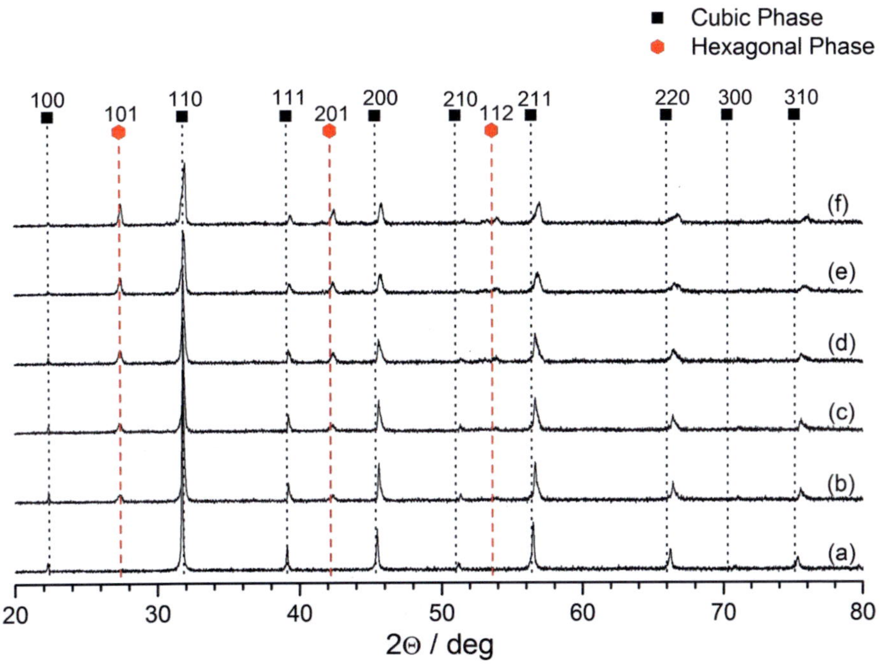

Fig. 8.10: XRD spectra of new BSCF powder (a) and after annealing at 800 °C in ambient air for 72 h (b), 120 h (c), 240 h (d), 600 h (e) and 1680 h (f), showing a continuous increase of the reflections from the hexagonal (101), (201) and (112) planes.

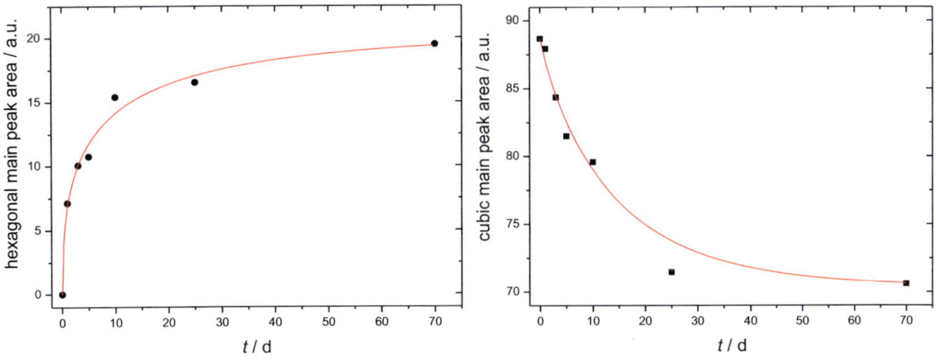

Fig. 8.11: Exponential fit to the area under the hexagonal (101) reflection (left), yielding a t_{90} of 45.5 days, and to the area under the cubic (110) reflection (right), yielding a t_{90} of 35.7 days. The corresponding fit parameters amounted to $\alpha_{HEX,eq} = 20.614 \pm 2.158$, $k = (0.407 \pm 0.051)\ \mathrm{d}^{-n}$, and $n = 0.454 \pm 0.091$ for the hexagonal reflection and $\alpha_{CUB,eq} = 70.342 \pm 1.329$, $k = (0.100 \pm 0.034)\ \mathrm{d}^{-n}$, and $n = 0.877 \pm 0.201$ for the cubic reflection. In the latter case, the data point at $t = 25$ d was weighted with 1/25.

Phase growth or diminution is fitted by an equation of the Johnson-Mehl-Avrami-Kolmogorow type [21], i.e.

$$\alpha_{HEX}(t) = \alpha_{HEX,eq}\left(1 - e^{-kt^n}\right), \tag{9.6}$$

for the growth of the hexagonal phase, where α_{HEX} is the area of the hexagonal (101) peak and $\alpha_{HEX,eq}$ is the area after reaching phase equilibrium for $t \to \infty$, and

$$\alpha_{CUB}(t) = \alpha_{CUB,eq} + \alpha_{dif}e^{-kt^n} \quad , \tag{9.7}$$

for the diminution of the cubic phase, where $\alpha_{CUB,,eq}$ is the area of the cubic (110) peak for $t \to \infty$ and $\alpha_{dif} = \alpha_{CUB}(0) - \alpha_{CUB,eq}$. All peak areas are given in arbitrary units.

The obtained fit parameters amounted to $\alpha_{HEX,eq} = 20.614 \pm 2.158$, $k = (0.407 \pm 0.051)$ d^{-n} and $n = 0.454 \pm 0.091$, and $\alpha_{CUB,eq} = 70.342 \pm 1.329$, $k = (0.100 \pm 0.034)$ d^{-n} and $n = 0.877 \pm 0.201$, respectively. These fits yielded t_{90} values (the time after which 90 % of the mathematical limit of the exponential function is reached), of 45.5 days for the growth of the hexagonal phase and 35.7 days for the diminution of the cubic phase, thus showing this transition process to be as slow as previously reported by [14].

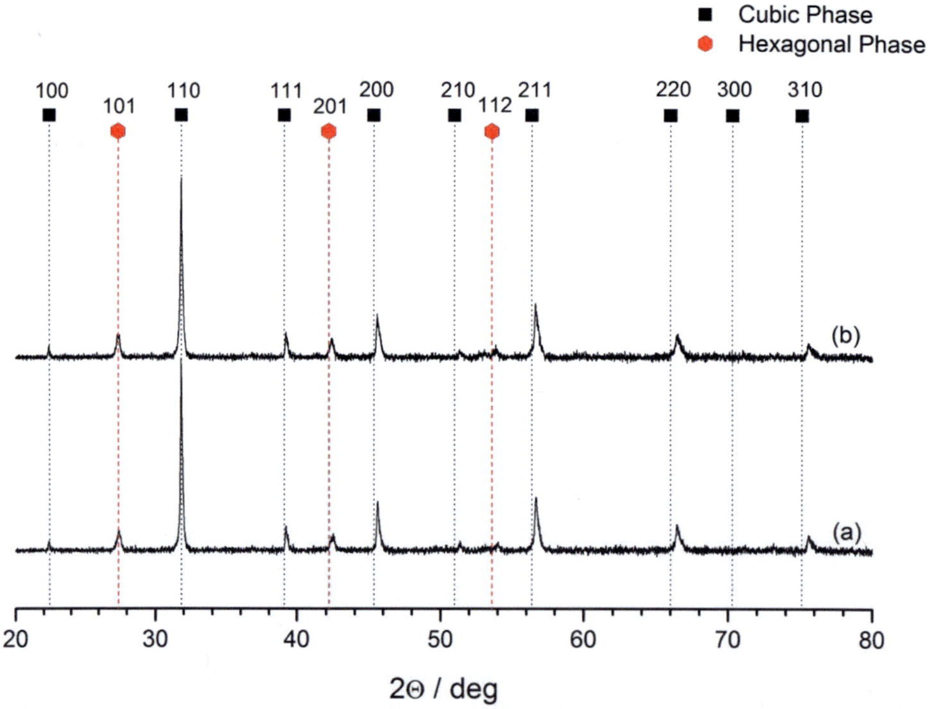

Fig. 8.12: XRD spectra of BSCF powder: annealing for 10 days at 800 °C in synthetic air (a) yields the same hexagonal phase growth as in ambient air (b), proving that it is not depending on other ambient gases.

The fairly large discrepancy between the fit parameters of the hexagonal and cubic phase changes may be attributed to the low amount and scatter of data points. However, it is obvious that there is a coexistence of both cubic and hexagonal phase at 800 °C, but this equilibrium will only be reached after many days.

To determine that this phase transition is not triggered by reactions with trace gases in the atmosphere, as for example the formation of carbonates with ambient CO_2 for which BSCF is known to be prone [6-9], Fig. 8.12 shows a comparison of the XRD spectra from powders annealed for 240 hours in synthetic air consisting of 20 % oxygen in nitrogen (a) and ambient air (b). Within the measurement accuracy of XRD, no difference between the spectra can be observed.

8.3.2.2 Temperature Limit for the Formation of the Hexagonal Phase

In order to further refine the reported temperature limit under which the cubic BSCF system will transform to the hexagonal structure [12], Fig. 8.13 shows the XRD spectra of several single phase cubic powders (a) after annealing for 72 hours in ambient air at 800 °C (b), 825 °C (c), 840 °C (d) and 850 °C (e). The formation of the hexagonal phase is determined to occur at temperatures below 840 °C.

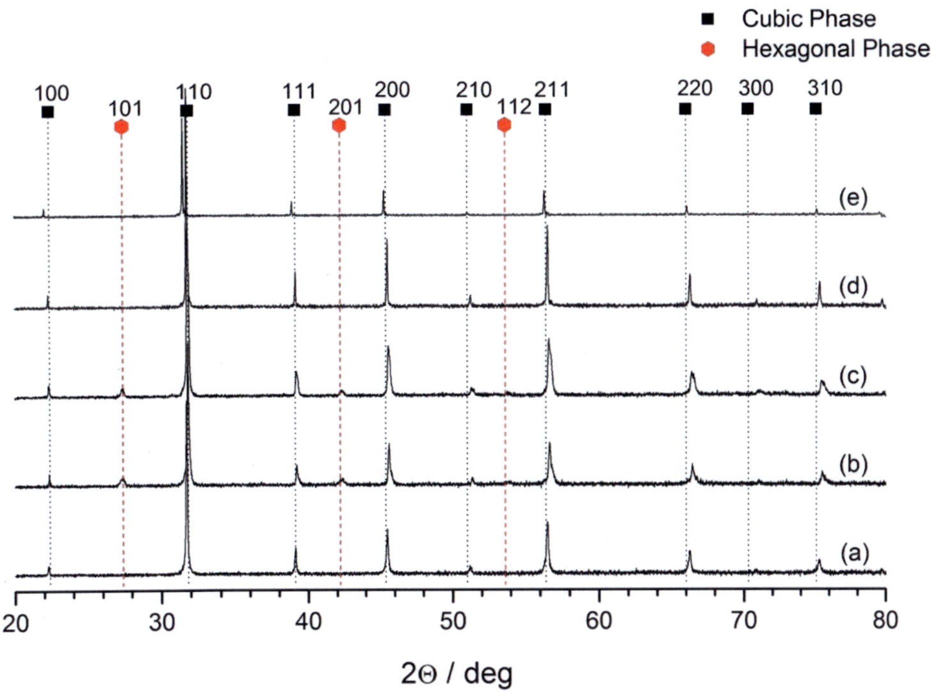

Fig. 8.13: XRD spectra of new BSCF powder (a) and after annealing for 72 hours in ambient air at 800 °C (b), 825 °C (c), 840 °C (d) and 850 °C (e). The hexagonal phase starts to form at temperatures below 840 °C.

8.3.2.3 Hexagonal Phase at 700 °C

To clarify the measured increase of the electrical conductivity after cooling to 700 °C from 800 °C in the long term measurement shown in Fig. 8.8, fresh BSCF powder was annealed for 72 hours at 800 °C and then for further 72 hours at 700 °C. As can be seen in Fig. 8.14, the initially cubic perovskite (a) transforms partially to the hexagonal phase after the first treatment (b), the amount of which decreases again after treatment at the lower temperature (c).

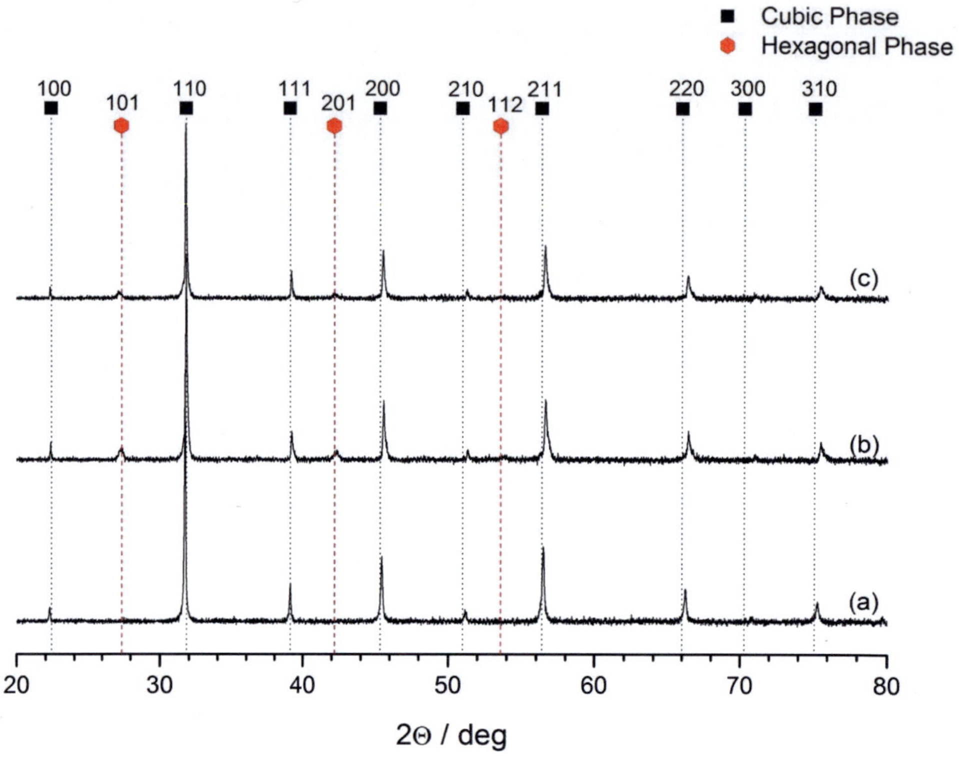

Fig. 8.14: XRD spectra of BSCF powder: the reflection of the hexagonal phase in powder annealed for 3 days at 800 °C (b) weakens after a subsequent annealing at 700 °C (c), indicating a different equilibrium of cubic and hexagonal phase at the lower temperature. For comparison, (a) shows again the spectrum of new powder.

Whether this indicates just a different phase equilibrium between cubic and hexagonal phase at these two different temperatures or the hexagonal phase would completely reform back to the cubic structure at 700 °C given time, has been examined by annealing pure BSCF cubic phase powder for 240 hours at that temperature. The comparison of the obtained XRD spectrum (a) to the one from after the 800 °C treatment (b) is shown in Fig. 8.15 and clearly proves the formation and stability of the hexagonal phase from the purely cubic configuration also at temperatures of 700 °C, meaning the coexisting of both phases also at this lower temperature with a different equilibrium shifted back more in favor of the cubic structure.

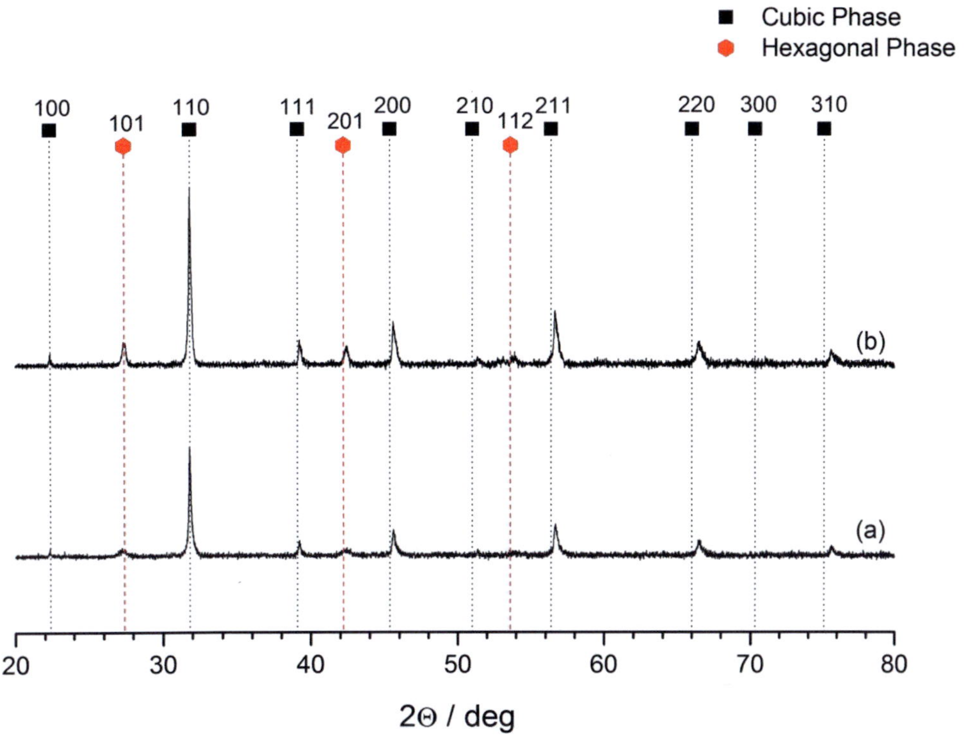

Fig. 8.15: XRD spectra of new single phase cubic BSCF powder after annealing for 10 days at 700 °C (a) and 800 °C (b), respectively. The formation of the hexagonal phase at 700 °C can be confirmed.

Analysis of Figure 8.14

	100	101	110	111	201	200	210	112	211	220	300	310	Sum	cub/hex
(b) cubic	22.02		549.93	80.19		192.24	29.17		278.54	138.59	13.83	86.92	1391.42	10.06
(b) hex		54.74			54.40			29.24					138.37	
(c) cubic	14.37		2690.72	62.26		143.08	21.34		204.21	100.56	16.16	67.09	3319.78	45.95
(c) hex		30.74			27.15			14.36					72.24	

Tab. 8.1: Integrated reflected X-ray intensities of the measurements shown in Fig. 8.14.

Table 8.1 lists the integrated intensities of the cubic and hexagonal reflections in Fig. 8.14, table 8.2 lists the corresponding data extracted from the measurement shown in Fig. 8.15. In those tables, the fractions of cubic and hexagonal peak areas are calculated to further illustrate this behavior. The fraction shifts from 10.06 back to 45.95 – therefore indicating a shift in the phase equilibrium towards the cubic lattice – after the subsequent 72 hours annealing at 700 °C shown in Fig 8.14. For the 240 hours annealing of fresh powder, the fraction calculates to 6.15 and 5.80 after treatment at 700 °C and 800 °C, respectively.

Analysis of Figure 8.15

	100	101	110	111	201	200	210	112	211	220	300	310	Sum	cub/hex
(b) cubic	17.90		527.12	65.78		167.96	24.41		261.06	148.44	12.61	83.70	1308.98	5.80
(b) hex		100.90			83.07			41.53					225.50	
(a) cubic	10.87		352.67	43.61		116.16	12.08		168.66	85.56	0.00	42.34	831.93	6.15
(a) hex		54.78			56.03			24.40					135.20	

Tab. 8.2: Integrated reflected X-ray intensities of the measurements shown in Fig. 8.15.

8.3.2.4 Reversibility of the Phase Transition

In the powder that has been annealed for 1680 hours at 800 °C, it is to be assumed that the phase transition nearly reached the specific equilibrium for that temperature.

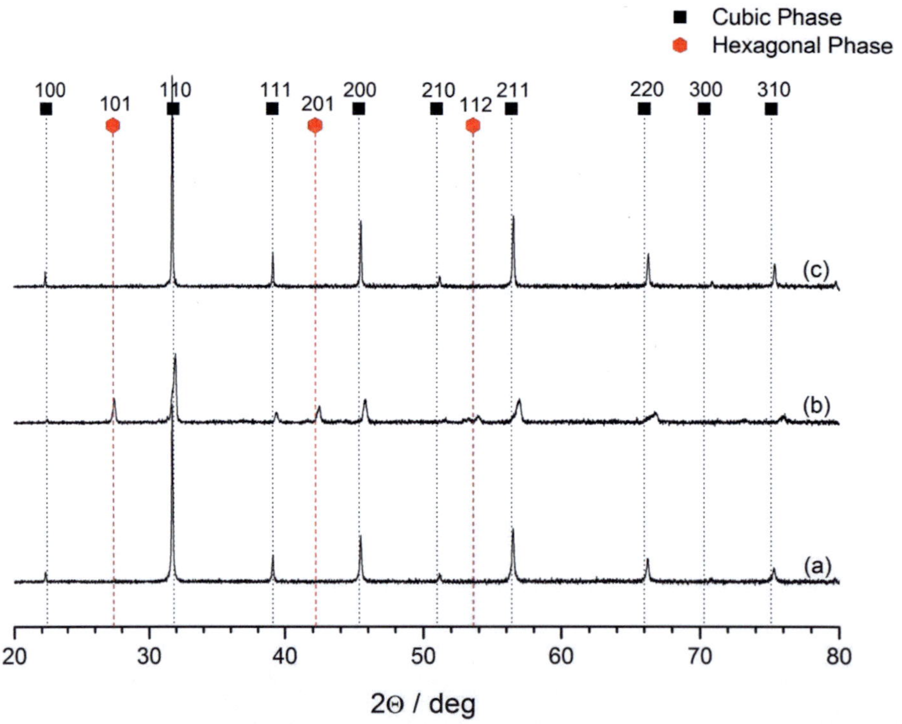

Fig. 8.16: XRD spectra of BSCF powder: After annealing for 72 hours at 900 °C (c), the hexagonal phase that formed during 1680 hours at 800 °C (b) reverts back to the cubic structure as in the fresh powder (a).

As earlier calcination experiments had shown that a stoichiometrically correct mixture of carbonates and oxides will form purely cubic BSCF after 30 hours at 900 °C (cf. Chapter 2), it could be argued that the hexagonal phase will reform to the cubic structure at this temperature. To this end, a further annealing of the dual phase powder at 900 °C was performed. Fig. 8.16 shows a comparison of the purely cubic fresh powder (a) with the dual phase BSCF containing a significant amount of the hexagonal configuration after 1680 hours at 800 °C (b) and the powder after subsequent treatment for 72 hours at 900 °C (c) which has reverted to consisting of only the cubic lattice again. Therefore, for BSCF powder it has been shown that the very slow phase transition at 800 °C is fully reversible in a short amount of time by temperature treatment at 900 °C.

8.4 Occurrence of Hexagonal Phase in BSCF Bulks

8.4.1 Long Term Conductivity Measurements

As the long term measurement discussed in section 8.3.1 suggests, the phase transition in dense ceramics can be observed by electrical conductivity measurements. The question remains if the change happens predominantly at the grain boundaries or through nucleation inside the matrix structure of the grains. The conductivities of three dense BSCF bulk ceramics with different average grain sizes have been measured at the same time for more than 1300 hours at 800 ºC in ambient air. Fig. 8.17 shows the normalized conductivities for the bulks with grain sizes of 25 µm (a), 50 µm (b) and 90 µm (c). Normalization to the time $t = 4$ h, after the furnace completed heating up, has been done for better comparison between the degradation behavior. Beforehand, all three measurements have been corrected for ambient air pressure changes[4] as described in section 8.3.1.

The best correction could be achieved for the bulk with the smallest grain size of 25 µm, as can be seen by the smaller impact the pressure drops around 450 hours and 600 hours into the measurement have on the conductivity for this sample (a) in comparison to the other ones (b, c). This can be attributed to the fact that the $\sigma(pO_2)$ dependency this correction bases upon has been determined through measurements on an identical bulk with a similar average grain size of 25 µm – also hinting at slightly different dependency of σ on pO_2 for different sized grains in the examined sample.

[4] Meteorological data were kindly provided by the Institute for Meteorology and Climate Research (IMK), KIT, Karlsruhe/Germany.

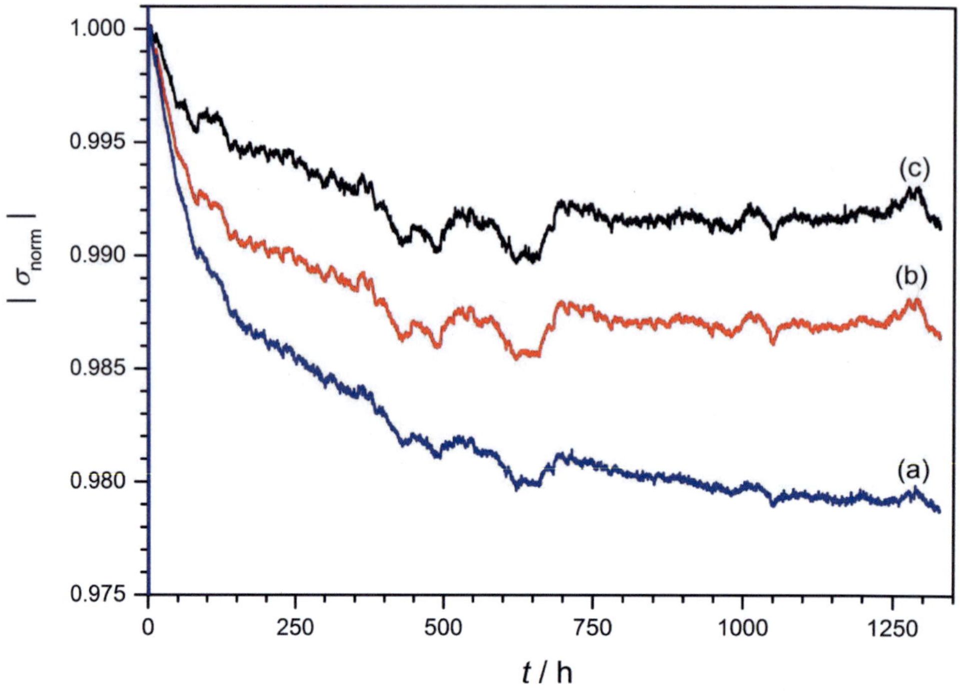

Fig. 8.17: Long term conductivity measurements at 800 °C on BSCF bulk ceramics (thickness $d = 1$ mm) with different medium grain sizes of 25 µm (B-BSCF-001-A, (a)), 50 µm (B-BSCF-004-A, (b)) and 90 µm (B-BSCF-003-A, (c)) show the highest degradation of the electrical conductivity for the bulk with the smallest grains, indicating that the formation of the hexagonal phase is occurring predominantly at the grain boundaries.

As the degradation in electrical conductivity is more pronounced the smaller the average grain size was, it is reasonable to assume that the formation of the hexagonal phase in bulk ceramics happens predominantly at the grain boundaries, which will have a greater impact on the electrical conductivity in samples with smaller grains. This is in agreement with the findings of Müller et al. [14] and the results shown in [2] and in Sections 8.4.2 and 8.5 of this Chapter.

Interestingly, the degradation behavior exhibited by these bulks after heating from room temperature directly to 800 °C seems to be different than that shown after cooling down from 900 °C shown in Fig. 8.8. While that decrease in conductivity appears to be nearly linear – on a longer timescale exponential behavior should be visible – in the measurement at hand there could be two different degradation processes involved, one occurring mainly throughout the first 100 hours, best visible in curve (a), followed by the slow degradation over the remaining measurement time.

SEM/TEM analyses performed by P. Müller – and published in our joint paper [2] – suggest that the formation of the plate-like phase discussed in the following part of this chapter occurs very fast (even after only 10 hours of annealing), while the powder diffractograms have shown a much slower formation of the hexagonal phase. This can explain the observed two-step degradation.

8.4.2 SEM/TEM Investigations on Bulk Samples

In the first part of this chapter it has been shown that the formation of a hexagonal phase at temperatures below 840 °C [1;3;12-16] is responsible for degradation of the electronic conductivity and inherent membrane performance. However, the timescales of the hexagonal phase formation in BSCF powders differ from those of the changes in sintered bulk sample conductivity – phase equilibration in powders is nearly achieved after approx. 40 days, while conductivity values still are subject to change after a similar time. It was to be expected that sintered samples will show slightly different behavior and therefore a detailed structural investigation on sintered samples becomes prudent, especially because any deviation from the cubic perovskite phase of BSCF causes a significant reduction of oxygen-ion conductivity [22]. Understanding the decomposition processes and a subsequent stabilization of the cubic system is of high importance for further membrane developments, especially in regard to producing dense BSCF thin films, considering their outstanding oxygen permeation reported in [23].

Limited success in stabilizing the cubic BSCF phase has been reported. The stability could be improved to some extent by doping with single-valent cations like Zr and Y [24-26]. Also, new promising materials were found in the search for cobalt-free high-performing membranes by exchanging Co with Zn [27-30], Al [31] or Cu [32]. However, further improvement of the material's thermodynamic stability necessitates a detailed understanding of the degradation processes on a microscopic scale.

In the first part of this chapter, an understanding of the phase changes in BSCF has been derived from correlating XRD measurements on powders and conductivity measurements on dense sintered samples. However, XRD only yields information on the phases without further details on microstructure, and phase changes in powders don't necessarily have to be identical to those occurring in sintered samples. So far, only few electron microscopic investigations of the microstructural changes and any correlation with the degradation processes have been published. The formation of the hexagonal phase has been investigated by Arnold et al. [13] and Mueller et al. [14]. However, it was found in our collaboration with the Laboratory for Electron Microscopy as well as by other groups [3;16;33] that at least one further phase with a plate-like morphology is involved in the degradation process. A new ordered lamellar phase with a 15R structure was found by Efimov et al. [16] and the similarity to $Ba_3Co_2O_9(Co_8O_8)$ reported by Sun et al. [34] was described. During this cooperation it could also be shown that the formation kinetics of the plate-like regions is fast compared to the formation of the hexagonal phase [3]. It therefore seems to suggest that these plate-like regions are at least partially responsible for the degradation of the electronic – as shown in Sections 8.3.1 and 8.4.1 – and oxygen permeation flux which was also reported to be faster at the beginning of long-term measurements [10]. This influence on material performance motivates further investigation of the plate-like regions.

In this section, results of detailed phase analyses of plate-like regions in BSCF by combining high-resolution (scanning) transmission electron microscopy (TEM, STEM) and chemical analyses by energy dispersive X-ray spectroscopy (EDX) performed at LEM by P. Müller and M. Meffert are shown. By analyzing the crystal structure and composition, considerable variations in structure and composition could be determined in plate-like regions. On the basis of the phase analyses, a detailed understanding of the formation mechanism of the plate-like regions and their influence on the degradation of BSCF has been achieved and published in [2].

8.4.2.1 Results

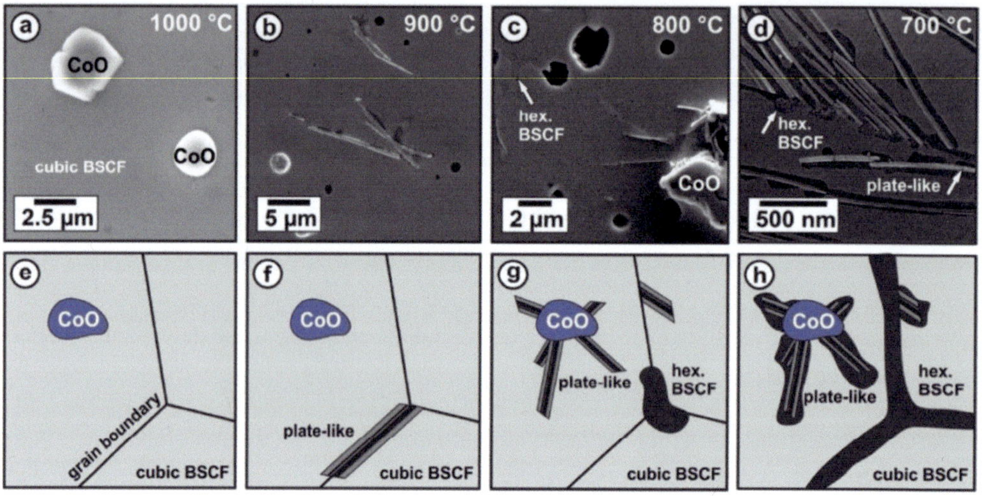

Fig. 8.18: Overview of detected phases present in BSCF samples after annealing in air for 100 h at 1000 °C, 900 °C, 800 °C and 700 °C, respectively (from left to right, B-BSCF-026A, B-BSCF-028A, B-BSCF-028B, and B-BSCF-029A). Images by P. Müller, LEM, KIT, from our joint publication [2].

In Fig. 8.18, images (a-d) show typical SEM micrographs of BSCF after annealing at temperatures between 700 °C and 1000 °C for 100 h and subsequent quenching in water as described in Section 8.2.4. Different phases can be distinguished by the different image intensities (different grey-scale values). As described in our joint paper [2], this can be achieved by a topographic phase contrast due to chemically etching the polished sample using colloidal silicon. In [3] it was established via TEM crystallography, that the intermediate intensity represents the cubic phase and dark areas depict the hexagonal phase, while plate-like regions appear with bright contrast and in their elongated shape.

As the depicted SEM images cannot show all microstructural features over a larger sample area, Figs. 8.18(e-h) schematically illustrate the summarized findings of numerous SEM images and TEM investigations of the grain boundaries.

The sample after annealing at 1000 °C, shown in Figs. 8.18(a,e), does not contain any hexagonal phase, only cubic BSCF is present, together with large particles (several µm in diameter) with bright SEM contrast that depict CoO precipitates.

After annealing at 900 °C (Figs. 8.18(b,f)), the presence of large plate-like regions (several µm in length) along the grain boundaries can be detected. Annealing at 800 °C (Figs. 8.18(c,g)) results in formation of the hexagonal phase, in agreement with the findings within the powder samples (cf. Section 8.3.2.2 and [1]), in addition to the plate-like regions. Precipitation of the hexagonal phase typically occurs at triple points of grain boundaries like the hexagonal region marked in Fig. 8.18(c). Also, at this temperature, smaller plate-like regions grow around CoO particles in addition to the formation along grain boundaries. At 700 °C (Figs. 8.18(d,h)), even more plate-like regions can be found, arranged in fishbone-like structures, and the hexagonal phase decorates nearly all grain boundaries and forms around present CoO particles as well as around the plate like regions.

In [2], we showed that TEM pictures of typical plate-like regions after annealing at 700 °C reveal a lamellar substructure of these plates (Fig. 8.19(a)), while generally being surrounded by the hexagonal phase as shown in Figs. 8.18(d,h). HRTEM imaging of the interface between two lamellae (Fig. 8.19(b)) show them to consist of either the cubic phase or the hexagonal one, with the $(111)_{cubic}$ and $(0001)_{hexagonal}$ planes oriented in parallel and – measured in a diffractogram – their elementary cells rotated by 34° when viewed in the direction of the cubic [101] and hexagonal $[2\bar{1}\bar{1}0]$.

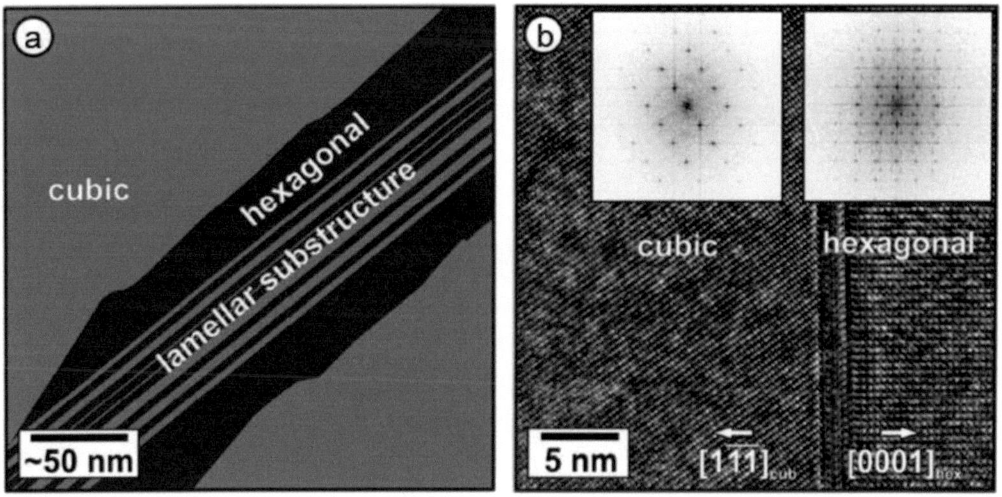

Fig. 8.19: (a) scheme of the microstructure of plate-like regions, (b) high-resolution TEM image of an interface in a plate-like region of sample B-BSCF-029A, containing a cubic and a hexagonal lamella with corresponding diffractograms (insets). Images by P. Müller, LEM, KIT, from our joint publication [2].

In the perovskite structure, the oxygen ion octahedrons are aligned corner-sharing in the cubic phase and face-sharing in the hexagonal phase. In the lattice direction mentioned above and observed in [2], the cubic (010)-planes and hexagonal ($01\bar{1}0$)-planes form an angle of 35.26° at the interface. This matches well with the measured rotation of 34°.

During detailed investigations, another phase with a different structure and composition was identified. It could be compared to structure models derived from a phase described by Sun et al. [34] and David et al. [35] and is also similar to the R15 phase described by Efimov et al. [16]. Sun et al. give its composition to be $Ba_3Co_{10}O_{17}$, therefore this phase has been denoted as BCO-type phase and described in detail in [2]. It has also been found as lamella beside cubic and hexagonal lamellae in plate-like regions.

The composition of the secondary phases in BSCF – that is plate-like regions with a random arrangement of cubic, hexagonal and BCO-type lamellae after annealing temperatures ≤ 900 °C as well as the hexagonal phase below 840 °C [3;15] – has been studied via extensive chemical analyses by standard-based EDX. Line-scans with one full EDX spectrum per pixel over regions with different phases were carried out in the FEI Titan operated in STEM mode. All spectra were then quantified with the Cliff-Lorimer factors as described in [2]. Table 8.2 contains the different chemical compositions with standard deviations in at%, obtained by averaging up to 200 spectra. Due to additional systematic errors resulting from imprecise background subtraction, peak-fitting and absorption effects as well as potential artifacts introduced by the Ar^+-ion milling, the overall systematic error is estimated to be ± 3 at%.

Phase	Ba [at%]	Sr [at%]	Co [at%]	Fe [at%]
$Ba_{0.5}Sr_{0.5}Co_{0.8}Fe_{0.2}O_{3-\delta}$	25	25	40	10
Cubic phase	26 ± 1	25.5 ± 1.5	40 ± 1.5	8.5 ± 1
Hexagonal phase	30 ± 1	27 ± 1.5	42 ± 1	1 ± 0.5
BCO-type phase	21.5 ± 3	3 ± 1	67.5 ± 2.5	8 ± 1

Tab. 8.2: Concentrations of Ba, Sr, Co and Fe in at% for the different observed BSCF phases determined by EDX compared to the nominal (intended) composition of cubic BSCF (first line). Measurements performed by P. Müller, LEM, KIT, from our joint publication [2].

All phases can easily be distinguished by their chemical composition. The cubic phase only shows small deviations from the nominal composition of $Ba_{0.5}Sr_{0.5}Co_{0.8}Fe_{0.2}O_{3-\delta}$. The hexagonal phase exhibits the absence of Fe while the chemical composition is still close to the ABO_3 stoichiometry of the perovskite structure. The BCO-type phases' composition strongly deviates from the other two phases with respect to the A/B-ratio due to a strongly enriched Co-content and the absence of Sr. In plate-like regions consisting of lamellae of all three phases (cubic, hexagonal, and BCO-type) the average chemical composition depends on the volume fraction of the different phases and therefore significant variations in average stoichiometry are detected.

As the hexagonal phase only forms at temperatures below 840 °C, the BCO-type phase seems to form the main portion of the plate-like regions at 900 °C. This also explains the strong Sr-depletion in the plate-like regions found in samples annealed above 840 °C [3] as well as their different microstructure at 900 °C (few large plate-like regions) and below 840 °C (many small plate-like regions). At temperatures below the formation threshold of the hexagonal phase all three phases are observed in plate-like regions. The influence of the plate-like regions on the degradation of the ionic conductivity may differ significantly from the hexagonal bulk phase which is discussed in the following.

While the hexagonal phases' influence on the degradation has already been discussed in earlier studies [12-14;16;36], its different chemical composition in comparison with the cubic phase has not been of particular interest. The B-sites in the face-sharing octahedrons of the hexagonal phase are more densely packed compared to the cubic BSCF phase. Therefore the B-site cations need to be smaller in order to fit into the structure. Cobalt remains in the hexagonal phase due to its ability to decrease its size by changing its valence state from 2+ to 3+ [16], whereas iron does not form smaller ions in BSCF [37], and therefore has to be expelled from the hexagonal phase. In [2] we confirm this by detecting Fe-cation excess and Co-cation depletion in the vicinity of the hexagonal phase via EDX linescans.

Due to this required cation demixing, the formation of the hexagonal phase in BSCF occurs with the slow kinetics observed in Section 8.3 and by Mueller et al. [14]. A reduction in oxygen ion conductivity is caused by the reduction in oxygen nonstoichiometry because of the increased B-site valence state.

Plate-like regions form at a fast rate compared to bulk hexagonal phase [3] which accounts for the faster degradation during the early stages of long term measurements as shown in Section 8.4.1 and reported in literature [6;10]. Due to their composition of lamellae with hexagonal stacking and correspondingly reduced oxygen conductivity, the plate-like regions act as barriers for oxygen conduction. They also have been found to form predominantly at grain boundaries which are believed to contribute significantly to oxygen diffusion [33].

The BCO-type phase is expected to also contribute to the degradation of the ionic conductivity in BSCF due to its lower concentration of oxygen vacancies which constitutes a diffusion barrier.

8.5 Stabilization of the Cubic Perovskite Phase by B-site Doping

As the stability of the cubic phase could be improved to some extent by doping with single-valent cations like Zr and Y [24-26], sintered samples with various dopant concentrations of Zr, Y and Sc have been prepared and analyzed similar to the pure BSCF samples discussed in the above part of this chapter. Samples are designated as BSCFnX, where n denotes the dopant concentration in at% and X stands for the dopant itself.

Fig. 8.20 shows SEM micrographs of chemically etched samples of BSCF doped with 3 at% of Zr, Y or Sc in comparison to undoped BSCF after annealing for 100 hours at 700 °C. While the Zr proves not to prevent the formation of plate-like and hexagonal phase (Fig. 8.20(a)), the introduction of 3 at% of Y on the B-site in the BSCF matrix hinders the formation of plate-like regions and only the percolating hexagonal phase at the grain boundaries remains (Fig. 8.20(b)) The Sc-doped sample exhibits both secondary phases (Fig. 8.20(c)) that also occur in undoped BSCF (Fig. 8.20(d)).

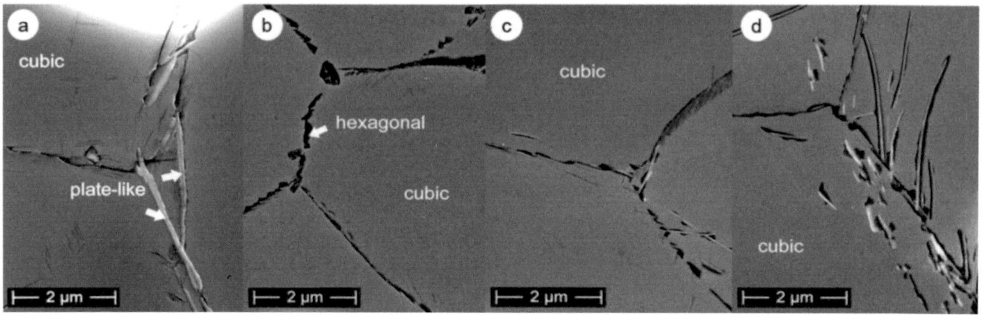

Fig. 8.20: SEM pictures of different phases in (a) BSCF3Zr (B-BSCF3Zr-151A), (b) BSCF3Y (B-BSCF3Y-163A), (c) BSCF3Sc (B-BSCF3Sc-157A), and (d) undoped BSCF (B-BSCF-029A) after annealing for 100 h at 700 °C and subsequent quenching. Images by M. Meffert, LEM, KIT, from our joint publication [38].

More examinations have been performed on similar bulks after annealing at temperatures of 800 °C and 900 °C and are schematically summarized in a similar fashion to Section 8.4.2.1 in Fig. 8.21.

In all analyzed samples it could be confirmed that the addition of 3 at% of Zr (a-c) as well as Sc (g-f) does neither prevent the formation of the plate-like phase even at the high temperature of 900 °C nor the development of the hexagonal phase at the lower temperatures of 800 °C and 700 °C. Secondary phase formation in bulks of these stoichiometries is quite similar to that in undoped BSCF as shown in Fig. 8.18.

Doping with 3 at% of Y however not only prevents the occurrence of CoO precipitates at all temperatures, it also proves to completely suppress any formation of the plate-like phase (d-f). Even though the hexagonal phase is still present, the absence of plate-like regions suggests a stabilized performance of this material in comparison to undoped BSCF.

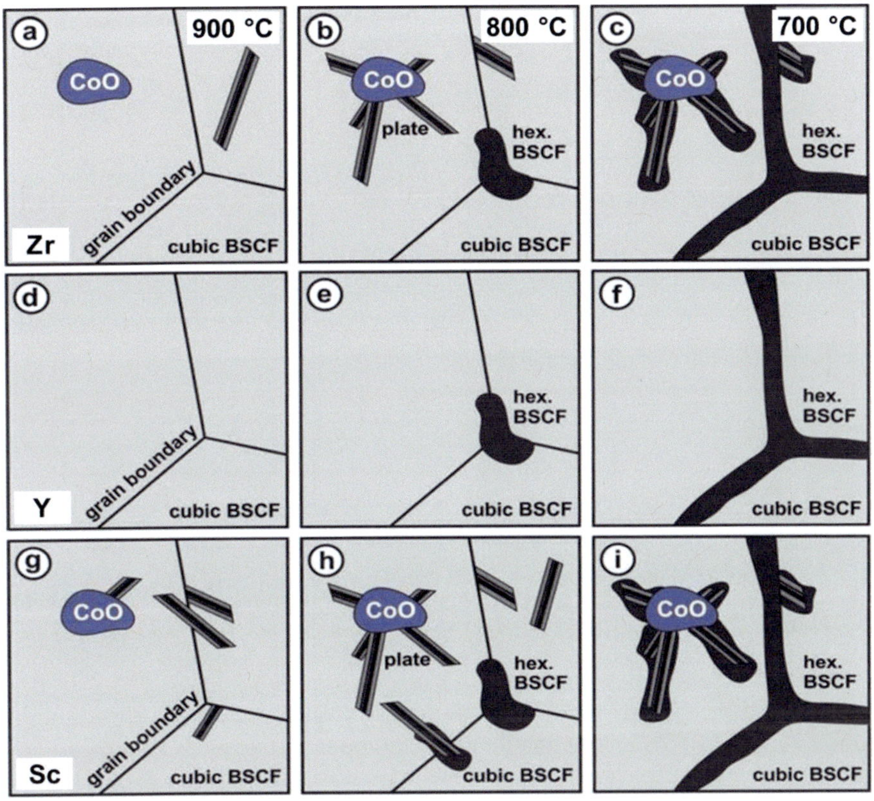

Fig. 8.21: Overview of detected phases present in BSCF3X samples after annealing in air for 100 hours at 900 °C, 800 °C, 700 °C, respectively (from left to right) for BSCF doped with 3 at% Zr, Y and Sc (top to bottom, respectively). Samples:
top row left to right: B-BSCF3Zr-205B, B-BSCF3Zr-207A, B-BSCF3Zr-151A; middle row left to right: B-BSCF3Y-161B, B-BSCF3Y-162A, B-BSCF3Y-163A; bottom row left to right: B-BSCF3Sc-155B, B-BSCF3Sc-156A, B-BSCF3Sc-157A. Images by P. Müller, LEM, KIT [39].

Therefore, a long term conductivity measurement has been carried out (Fig. 8.22), where sintered samples of undoped BSCF (black curve), BSCF3Y (red curve) and BSCF10Y (blue curve) were simultaneously measured for several hundreds of hours at 900 °C and 800 °C. The electrical conductivity of the samples decreases with increasing dopant content, which was to be expected (cf. Chapter 4), leaving the 10% Y-doped BSCF at approx. 21 S/cm at 900 °C, which is only half of the value for undoped BSCF. However, while the degradation is clearly visible for BSCF and, to a lesser extent, also for BSCF3Y, the BSCF10Y sample shows no degradation at all at both temperatures. In this case, no correction for the meteorological changes in ambient air pressure (cf. Section 8.3.1) has been performed in order to provide optimal comparability between the different materials.

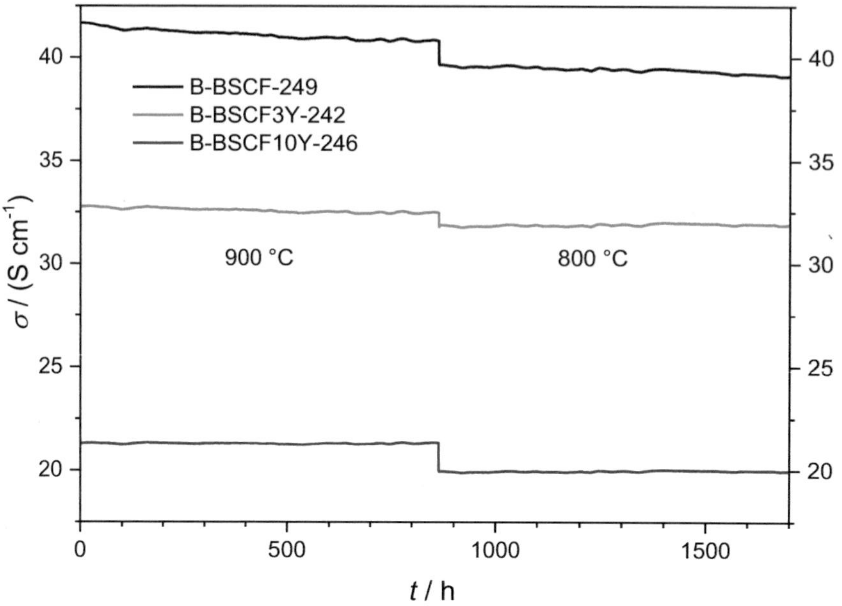

Fig. 8.22: Long term conductivity measurements for several hundreds of hours at 900 °C and 800 °C for BSCF, BSCF3Y and BSCF10Y bulk samples. The BSCF10Y sample shows stable conductivity values over the whole measurement.

The increasing content of Y prevents the formation of the plate like phase, due to the large ionic radius of the yttrium in comparison to iron, while both monovalent cations (for Fe in BSCF cf. [37]) would need to diffuse out of the oxygen octahedrons to enable their face sharing positions in the hexagonal structure. This hindered cation diffusion seems to stabilize the material system at the application-relevant temperature of 800 °C. The achievable oxygen flux through a membrane consisting of Y-doped BSCF has been measured for low dopant concentrations of 2.5 at% (as nominal substitution of the iron content in this case) by Haworth et al. [25] and found to be comparable to undoped BSCF within the uncertainties of the measurement concerning the absolute oxygen flux. For 10 at% Y content preliminary measurements by S. Baumann at the Forschungszentrum Jülich (Jülich, Germany) suggest a resulting overall drop in oxygen permeation of approx. 20 %. A long term oxygen flux measurement on our samples will be performed by S. Baumann in the near future.

By introducing small amounts of monovalent yttrium into the BSCF matrix, phase degradation – and subsequent degradation in electrical performance – could be prevented in the temperature range between 900 °C and 800 °C. This promises long term stable membrane performance at these highly relevant temperatures.

8.6 Conclusions

8.6.1 Powder X-ray Diffraction and Conductivity Measurements

The thermal stability of the cubic phase in $Ba_{0.5}Sr_{0.5}Co_{0.8}Fe_{0.2}O_{3-\delta}$ (BSCF) in the targeted temperature range for oxygen-transport membrane (OTM) applications (700...900 °C) has been investigated by means of an XRD analysis of powders, but for the first time in combination with electrical measurements performed on bulk samples of the same material. The XRD measurements on BSCF powders after annealing at different temperatures for various durations yielded the following results:

- In the single phase cubic BSCF system, a slow formation of the hexagonal phase occurs at $T = 800$ °C, leading to a phase equilibrium between cubic and hexagonal phase with a t_{90} of approx. 40 days.
- For annealing times of 72 h, phase transition is observable at $T < 840$ °C.
- The hexagonal phase also forms at $T = 700$ °C, albeit resulting in an equilibrium shifted more to the cubic phase than that at $T = 800$ °C.
- The phase transition is independent of any trace gases in the atmosphere, e.g. CO_2.
- The hexagonal phase transforms back to purely cubic after annealing at $T = 900$ °C.

The results of electrical measurements on BSCF bulk samples led to the following conclusions:

- In long term conductivity measurements on BSCF, the influence of ambient air pressure is clearly visible in the conductivity and can mostly be corrected by explicit knowledge of $\sigma(pO_2, T)$.
- While obtaining stable values over several hundred hours at $T = 900$ °C, a pronounced decrease in conductivity occurs over time at $T = 800$ °C, followed by an equally slow increase after a subsequent temperature reduction to 700 °C.
- This is in agreement with the formation of the hexagonal phase observed in BSCF powder at the temperatures in question, taking into account that the phase transition in dense ceramics is prone to be slower than in powder with a D_{50} of 2.40 µm.
- By heating new bulks directly to 800 °C, omitting the first step at 900 °C, a second, faster degradation process seems to be present in the conductivity measurement, pointing to another phase transformation occurring in the BSCF system, revealed to be the formation of plate-like regions consisting of a lamellar substructure as detected by SEM/TEM investigations by P. Müller at LEM, KIT (cf. second part of conclusions in Section 8.6.2.

- The degradation in conductivity is more pronounced in samples with smaller average grain size, suggesting the growth of the hexagonal phase to occur predominantly at the grain boundaries.

8.6.2 SEM/TEM Analyses

$Ba_{0.5}Sr_{0.5}Co_{0.8}Fe_{0.2}O_{3-\delta}$ bulk samples have been annealed at temperatures between 700 °C and 1000 °C (100 °C steps) for 100 hours each and subsequently quenched in water. Electron microscopic techniques and EDX were performed by P. Müller (LEM, KIT) in order to gain an overview of the formation of secondary phases from the cubic BSCF phase, which are correlated with the degradation in electrochemical performance summarized in Section 8.6.1.

- Treatment at 1000 °C yields only the cubic BSCF phase and CoO precipitates. For the 900 °C samples, large plate-like regions have been found at the grain boundaries. In the sample annealed at 800 °C, however, the hexagonal phase has been detected, typically at triple points of the grain boundaries, as well as smaller (in comparison to the 900 °C sample) plate-like regions at grain boundaries and CoO precipitates in larger amounts than in the 900 °C sample. Finally, at 700 °C, the growth of plate-like regions proved to be even more pronounced. The hexagonal phase often completely decorates grain boundaries and often surrounds plate-like regions.
- The plate-like regions exhibit a lamellar substructure: Lamellae of cubic and hexagonal phase alternate with an additional $Ba_{n+1}Co_nO_{3n+3}(Co_8O_8)$ (BCO)-type ($n \geq 2$) barium cobalt oxide phase.
- As the radius of iron cations is too large for them to occupy the B-sites, the hexagonal phase is strongly iron-depleted. The rate at which the hexagonal phase is formed is therefore limited by the interdiffusion of iron and cobalt.
- The oxygen transport in BSCF is detrimentally affected by the occurrence of these secondary phases. It is safe to assume that, in particular, the influence of the fast developing plate-like regions has so far not been sufficiently considered as being the main reason for pronounced performance degradation during the first several hundred hours of operation. This may be due to the impossibility of distinguishing this phase in a sample consisting of cubic and hexagonal phase by XRD measurements. These SEM/TEM analyses, however, can shed new light on the two-step degradation in electrical long-term conductivity measurements at 800 °C shown in Fig. 8.17, where a fast decrease was observed during the first 100 h, now attributed to the formation of the plate-like regions. At 700 °C and 800 °C, the plate-like phase – as well as the hexagonal phase – predominantly forms at grain boundaries, representing hindrances in respect to oxygen transport through the BSCF membrane.

- Yttrium doped BSCF shows no formation of the plate-like regions at all four temperatures (Fig. 8.21) and therefore no decrease in electrical conductivity occurs (Fig. 8.22). By this substitution, the BSCF system can be stabilized to operate at temperatures lower than 900 °C; therefore increasing the efficiency in an oxygen transport membrane module for air separation.

References

[1] C. Niedrig, S. Taufall, M. Burriel, W. Menesklou, S. F. Wagner, S. Baumann and E. Ivers-Tiffée, "Thermal stability of the cubic phase in $Ba_{0.5}Sr_{0.5}Co_{0.8}Fe_{0.2}O_{3-\delta}$ (BSCF)", *Solid State Ionics* **197**, pp. 25-31 (2011).

[2] P. Müller, H. Störmer, M. Meffert, L. Dieterle, C. Niedrig, S. F. Wagner, E. Ivers-Tiffée and D. Gerthsen, "Secondary phase formation in $Ba_{0.5}Sr_{0.5}Co_{0.8}Fe_{0.2}O_{3-d}$ studied by electron microscopy", *Chemistry of Materials* **25**, pp. 564-573 (2013).

[3] P. Müller, H. Störmer, L. Dieterle, C. Niedrig, E. Ivers-Tiffée and D. Gerthsen, "Decomposition pathway of cubic $Ba_{0.5}Sr_{0.5}Co_{0.8}Fe_{0.2}O_{3-\delta}$ between 700 °C and 1000 °C analyzed by electron microscopic techniques", *Solid State Ionics* **206**, pp. 57-66 (2012).

[4] P. Müller, H. Störmer, D. Gerthsen, J. Santiso, J. Roqueta, M. Burriel, C. Niedrig, S. F. Wagner and E. Ivers-Tiffée, "TEM investigations of mixed ionic/electronic conducting $Ba_{0.5}Sr_{0.5}Co_{0.8}Fe_{0.2}O_{3-\delta}$ materials", in W. Grogger, F. Hofer, and P. Pölt (Eds.), *Proceedings of the Microscopy Conference 2009*, Vol. **3**, Graz, Austria: Verlag der TU Graz, pp. 295-296 (2009).

[5] C. Niedrig, S. Taufall, P. Müller, S. F. Wagner, H. Störmer, D. Gerthsen and E. Ivers-Tiffée, "Thermal Stability of $Ba_{0.5}Sr_{0.5}Co_{0.8}Fe_{0.2}O_{3-\delta}$", in *First International Conference on Materials for Energy, Extended Abstracts - Book A*: Dechema, pp. 48-50 (2010).

[6] M. Arnold, H. Wang and A. Feldhoff, "Influence of CO_2 on the oxygen permeation performance and the microstructure of perovskite-type $(Ba_{0.5}Sr_{0.5})(Co_{0.8}Fe_{0.2})O_{3-\delta}$ membranes", *Journal of Membrane Science* **293**, pp. 44-52 (2007).

[7] A. Yan, M. Cheng, Y. L. Dong, W. S. Yang, V. Maragou, S. Q. Song and P. Tsiakaras, "Investigation of a $Ba_{0.5}Sr_{0.5}Co_{0.8}Fe_{0.2}O_{3-\delta}$ based cathode IT-SOFC I. The effect of CO_2 on the cell performance", *Applied Catalysis B-Environmental* **66**, pp. 64-71 (2006).

[8] E. Bucher, A. Egger, G. B. Caraman and W. Sitte, "Stability of the SOFC cathode material $(Ba,Sr)(Co,Fe)O_{3-\delta}$ in CO_2-containing atmospheres", *Journal of The Electrochemical Society* **155**, pp. B1218-B1224 (2008).

[9] A. Waindich, A. Möbius and M. Müller, "Corrosion of $Ba_{1-x}Sr_xCo_{1-y}Fe_yO_{3-\delta}$ and $La_{0.3}Ba_{0.7}Co_{0.2}Fe_{0.8}O_{3-\delta}$ materials for oxygen separating membranes under Oxycoal conditions", *Journal of Membrane Science* **337**, pp. 182-187 (2009).

[10] Z. P. Shao, W. S. Yang, Y. Cong, H. Dong, J. H. Tong and G. X. Xiong, "Investigation of the permeation behavior and stability of a $Ba_{0.5}Sr_{0.5}Co_{0.8}Fe_{0.2}O_{3-\delta}$ oxygen membrane", *Journal of Membrane Science* **172**, pp. 177-188 (2000).

[11] K. Efimov, T. Halfer, A. Kuhn, P. Heithans, J. Caro and A. Feldhoff, "Novel Cobalt-Free Oxygen-Permeable Perovskite-Type Membrane", *Chemistry of Materials* **22**, pp. 1540-1544 (2010).

[12] S. Švarcová, K. Wiik, J. Tolchard, H. J. M. Bouwmeester and T. Grande, "Structural instability of cubic perovskite $Ba_xSr_{1-x}Co_{1-y}Fe_yO_{3-\delta}$", *Solid State Ionics* **178**, pp. 1787-1791 (2008).

[13] M. Arnold, T. M. Gesing, J. Martynczuk and A. Feldhoff, "Correlation of the formation and the decomposition process of the BSCF perovskite at intermediate temperatures", *Chemistry of Materials* **20**, pp. 5851-5858 (2008).

[14] D. N. Müller, R. A. De Souza, T. E. Weirich, D. Roehrens, J. Mayer and M. Martin, "A kinetic study of the decomposition of the cubic perovskite-type oxide $Ba_xSr_{1-x}Co_{0.8}Fe_{0.2}O_{3-\delta}$ (BSCF) (x=0.1 and 0.5)", *Physical Chemistry Chemical Physics* **12**, pp. 10320-10328 (2010).

[15] P. Müller, L. Dieterle, E. Müller, H. Störmer, D. Gerthsen, C. Niedrig, S. Taufall, S. F. Wagner and E. Ivers-Tiffée, "$Ba_{0.5}Sr_{0.5}Co_{0.8}Fe_{0.2}O_{3-\delta}$ for Oxygen Separation Membranes", *ECS Transactions* **28**, pp. 309-314 (2010).

[16] K. Efimov, Q. Xu and A. Feldhoff, "Transmission Electron Microscopy Study of $Ba_{0.5}Sr_{0.5}Co_{0.8}Fe_{0.2}O_{3-\delta}$ Perovskite Decomposition at Intermediate Temperatures", *Chemistry of Materials* **22**, pp. 5866-5875 (2010).

[17] A. V. Crewe, "Scanning Electron Microscopes: Is High Resolution Possible?", *Science* **154**, pp. 729-738 (1966).

[18] A. V. Crewe, J. Wall and L. M. Welter, "A High-Resolution Scanning Transmission Electron Microscope", *Journal of Applied Physics* **39**, pp. 5861-5868 (1968).

[19] J. Wall, J. Langmore, M. Isaacson and A. V. Crewe, "Scanning Transmission Electron Microscopy at High Resolution", *Proceedings of the National Academy of Sciences of the United States of America* **71**, pp. 1-5 (1974).

[20] M. Burriel, C. Niedrig, W. Menesklou, S. F. Wagner, J. Santiso and E. Ivers-Tiffée, "BSCF epitaxial thin films: Electrical transport and oxygen surface exchange", *Solid State Ionics* **181**, pp. 602-608 (2010).

[21] J. W. Christian, *The theory of transformations in metals and alloys*, 3 ed., Oxford, New York: Pergamon Press (2002).

[22] M. Mogensen, D. Lybye, N. Bonanos, P. V. Hendriksen and F. W. Poulsen, "Factors controlling the oxide ion conductivity of fluorite and perovskite structured oxides", *Solid State Ionics* **174**, pp. 279-286 (2004).

[23] S. Baumann, J. M. Serra, M. P. Lobera, S. Escolástico, F. Schulze-Küppers and W. A. Meulenberg, "Ultrahigh oxygen permeation flux through supported $Ba_{0.5}Sr_{0.5}Co_{0.8}Fe_{0.2}O_{3-\delta}$ membranes", *Journal of Membrane Science* **377**, pp. 198-205 (2011).

[24] S. Yakovlev, C.-Y. Yoo, S. Fang and H. J. M. Bouwmeester, "Phase transformation and oxygen equilibration kinetics of pure and Zr-doped $Ba_{0.5}Sr_{0.5}Co_{0.8}Fe_{0.2}O_{3-\delta}$ perovskite oxide probed by electrical conductivity relaxation", *Applied Physics Letters* **96**, pp. 1-3 (2010).

[25] P. Haworth, S. Smart, J. Glasscock and J. C. Diniz da Costa, "Yttrium doped BSCF membranes for oxygen separation", *Separation and Purification Technology* **81**, pp. 88-93 (2011).

[26] P. Haworth, S. Smart, J. Glasscock and J. C. Diniz da Costa, "High performance yttrium-doped BSCF hollow fibre membranes", *Separation and Purification Technology* **94**, pp. 16-22 (2012).

[27] H. Wang, C. Tablet, A. Feldhoff and J. Caro, "A Cobalt-Free Oxygen-Permeable Membrane Based on the Perovskite-Type Oxide $Ba_{0.5}Sr_{0.5}Zn_{0.2}Fe_{0.8}O_{3-\delta}$", *Advanced Materials* **17**, pp. 1785-1788 (2005).

[28] A. Feldhoff, J. Martynczuk and H. Wang, "Advanced $Ba_{0.5}Sr_{0.5}Zn_{0.2}Fe_{0.8}O_{3-\delta}$ perovskite-type ceramics as oxygen selective membranes: Evaluation of the synthetic process", *Progress in Solid State Chemistry* **35**, pp. 339-353 (2007).

[29] A. Feldhoff, J. Martynczuk, M. Arnold, M. Myndyk, I. Bergmann, V. Šepelák, W. Gruner, U. Vogt, A. Hähnel and J. Woltersdorf, "Spin-state transition of iron in $(Ba_{0.5}Sr_{0.5})(Fe_{0.8}Zn_{0.2})O_{3-\delta}$ perovskite", *Journal of Solid State Chemistry* **182**, pp. 2961-2971 (2009).

[30] J. Martynczuk, K. Efimov, L. Robben and A. Feldhoff, "Performance of zinc-doped perovskite-type membranes at intermediate temperatures for long-term oxygen permeation and under a carbon dioxide atmosphere", *Journal of Membrane Science* **344**, pp. 62-70 (2009).

[31] J. Martynczuk, F. Liang, M. Arnold, V. Šepelák and A. Feldhoff, "Aluminum-Doped Perovskites As High-Performance Oxygen Permeation Materials", *Chemistry of Materials* **21**, pp. 1586-1594 (2009).

[32] L. Zhao, B. He, X. Zhang, R. Peng, G. Meng and X. Liu, "Electrochemical performance of novel cobalt-free oxide $Ba_{0.5}Sr_{0.5}Fe_{0.8}Cu_{0.2}O_{3-\delta}$ for solid oxide fuel cell cathode", *Journal of Power Sources* **195**, pp. 1859-1861 (2010).

[33] F. Liang, H. Jiang, H. Luo, J. Caro and A. Feldhoff, "Phase Stability and Permeation Behavior of a Dead-End $Ba_{0.5}Sr_{0.5}Co_{0.8}Fe_{0.2}O_{3-\delta}$ Tube Membrane in High-Purity Oxygen Production", *Chemistry of Materials* **23**, pp. 4765-4772 (2011).

[34] J. Sun, M. Yang, G. Li, T. Yang, F. Liao, Y. Wang, M. Xiong and J. Lin, "New Barium Cobaltite Series $Ba_{n+1}Co_nO_{3n+3}(Co_8O_8)$: Intergrowth Structure Containing Perovskite and CdI_2-Type Layers", *Inorganic Chemistry* **45**, pp. 9151-9153 (2006).

[35] R. David, A. Pautrat, H. Kabbour, M. Sturza, S. Curelea, G. André, D. Pelloquin and O. Mentré, "$[BaCoO_3]_n[BaCo_8O_{11}]$ Modular Intergrowths: Singularity of the $n = 2$ Term", *Chemistry of Materials* **23**, pp. 5191-5199 (2011).

[36] B. Rutkowski, J. Malzbender, R. W. Steinbrech, T. Beck and H. J. M. Bouwmeester, "Influence of thermal history on the cubic-to-hexagonal phase transformation and creep behaviour of $Ba_{0.5}Sr_{0.5}Co_{0.8}Fe_{0.2}O_{3-\delta}$ ceramics", *Journal of Membrane Science* **381**, pp. 221-225 (2011).

[37] A. S. Harvey, F. J. Litterst, Z. Yang, J. L. M. Rupp, A. Infortuna and L. J. Gauckler, "Oxidation states of Co and Fe in $Ba_{1-x}Sr_xCo_{1-y}Fe_yO_{3-\delta}$ (x, y = 0.2-0.8) and oxygen desorption in the temperature range 300-1273 K", *Physical Chemistry Chemical Physics* **11**, pp. 3090-3098 (2009).

[38] M. Meffert, P. Müller, H. Störmer, C. Niedrig, S. F. Wagner, E. Ivers-Tiffée, and D. Gerthsen, *Electron microscopy study of* $(Ba_{0.5}Sr_{0.5})(Co_{0.8}Fe_{0.2})_{0.97}X_{0.03}O_{3-\delta}$ *(X = Zr, Y, Sc)*, 15th European Microscopy Congress (Manchester, UK), 16.09. - 21.09.2012

[39] P. Müller, *Electron Microscopical Investigations of Doped and Undoped* $Ba_{0.5}Sr_{0.5}Co_{0.8}Fe_{0.2}O_{3-d}$ *for Oxygen Separation Membranes*, Dissertation, Laboratorium für Elektronenmikroskopie (LEM), Karlsruher Institut für Technologie (KIT) (2013).

Chapter 9

Summary and Outlook

The pressing problem of growing anthropogenic CO_2-emissions into the atmosphere requires new methods in order to avoid a global climate catastrophe. Carbon Capture and Sequestration (CCS) technology is in active development, aiming to separate CO_2 from the exhaust gases of a combustion process with the intention of storing the greenhouse gas securely (for example in depleted natural gas reservoirs) before it can be introduced into the atmosphere. Prime candidates for the implementation of this concept are highly localized point sources of vast amounts of CO_2, mainly fossil fueled power plants around the world.

Amongst other concepts, the Oxyfuel process can provide nearly pure CO_2 after a combustion process by avoiding the introduction of nitrogen into the process. In principle, this means burning the fuel in a pure oxygen atmosphere instead of using ambient air, consequently receiving pure CO_2 in the exhaust gas after filtering soot, SO_x, and condensing the present water vapor.

Providing pure oxygen in the vast amounts needed in a power plant is an energy demanding task, resulting in severe reduction of power plant efficiency. The most effective way could be the implementation of ceramic oxygen transport membranes (OTMs). Mixed ionic-electronic conducting (MIEC) ceramics can – at elevated temperatures above 700 °C – provide the necessary amount of oxygen while needing only half of the energy that would be consumed by cryogenic air separation techniques.

As very promising candidates for membrane materials, perovskites ABO_3 (with A: Sr, Ba, La, … and B: Fe, Co, …) have been under special investigation by the scientific community due to their relatively good stability, flexibility concerning cation substitutions, and high achievable oxygen fluxes. Four most interesting materials already known as cathode materials from solid oxide fuel cell (SOFC) research are $La_{0.58}Sr_{0.4}Co_{0.2}Fe_{0.8}O_{3-\delta}$ (LSCF), $Pr_{0.58}Sr_{0.4}Co_{0.2}Fe_{0.8}O_{3-\delta}$ (PSCF), $La_{0.6}Sr_{0.4}CoO_{3-\delta}$ (LSC), and $Ba_{0.5}Sr_{0.5}Co_{0.8}Fe_{0.2}O_{3-\delta}$ (BSCF), which all feature relatively good thermal and chemical stability while providing high oxygen permeation rates. These four materials are investigated in this thesis, with the main focus predominantly on BSCF, which is the sole candidate so far able to exceed an oxygen permeation rate of 10 $ml_N\ cm^{-2}\ min^{-1}$, which has been formulated during the MEM-BRAIN project[1] as minimum rate in order to guarantee feasibility of the use of a membrane air separation unit in fossil fueled power plants.

The defining parameters of an OTM are the oxygen transport parameters that describe the oxygen surface exchange (k^{δ}) and the bulk diffusion of oxygen ions through the crystal lattice (D^{δ}) and give information about the performance of a candidate material. Aside from the sole capability of providing high oxygen fluxes, ceramic membranes also have to be able to cope with the harsh operating conditions: stability at high temperatures, chemical stability issues in contact with the present gas atmospheres at both sides of the membrane, low oxygen partial pressures pO_2, as well as the necessary long term stability under these operating conditions.

This thesis features investigations into these issues pertaining mainly to the BSCF system, but branching out towards similar measurements on LSCF, PSCF and LSC where a new contribution to the community's knowledge could be made.

The materials were investigated in the form of powders, dense sintered bulk samples as well as thin films, each giving different opportunities to investigate the perovskite systems. Powders have been used to produce sintered samples and discern the sintering behavior of the material, as well as (in combination with pre- and post-test X-ray diffraction measurements) for thermal stability investigations (phase stability) and determining the stability limits at low oxygen partial pressures. Sintered bulk samples yielded insights into electrical behavior in dependence on temperature and oxygen partial pressure and provided the possibility for oxygen exchange parameter extraction by means of electrical conductivity relaxation (ECR) measurements as well as for electrical long term measurements presenting the chance to monitor slow phase changes in the material. Also, thin films deposited on $NdGaO_3$ (in the case of BSCF epitaxial as well as polycrystalline) presented the possibility of investigating surface related processes, i.e. surface morphology changes with temperature or low oxygen partial pressures, exchange parameter extraction at intermediate temperature by means of ECR in the frequency domain or in-situ degradation measurements in low-concentration CO_2 atmospheres. The epitaxial thin films additionally facilitated investigations of the BSCF system in the absence of grain boundaries. However, stress and strain introduced by the substrate influence the measurement results and have to be taken into account.

One central measurement setup in this thesis has been the closed tubular zirconia "oxygen pump", formerly used at IWE for partial pressure dependent conductivity measurements on, e.g., $SrTiO_3$-based materials, that has been upgraded in order to be used for measurements on the highly conductive MIEC oxides. The setup has further been extensively enhanced and validated (with the help of well-known metal oxides) in order to perform both coulometric titration investigations

[1] Helmholtz Alliance MEM-BRAIN: "Gas separation membranes for zero-emission fossil power plants", joint project funded by the German Helmholtz Association of German Research Centers, duration: 2007-2011.

with respect to pO_2-dependent stability and oxygen stoichiometry changes as well as electrical conductivity relaxation (ECR) measurements on different MIEC compositions.

Once established, oxygen stoichiometry changes in BSCF could be quantified in dependence on oxygen partial pressure down to its decomposition limit. For all investigated compositions (LSCF, PSCF, BSCF, LSC) the pO_2 stability range at 700...900 °C is easily sufficient for most applications (SOFC cathodes, oxygen-transport membranes).

Measurements in this setup also presented an inherent problem pertaining to conductivity values of BSCF determined and published in literature at mid-range temperatures (400 °C...600 °C) and low oxygen partial pressures ($pO_2 < 10^{-4}$ bar): in both cases, true equilibrium conductivity values can only be extracted after – sometimes extensively – long dwelling times. Conductivity measurements over temperature reveal extensive hysteresis while heating up and cooling down, requiring dwelling times of several tens of hours to get the true equilibrium conductivity at a given mid-range temperature between 400 °C and 600 °C. The same problem occurs at low oxygen partial pressures, requiring – even though the temperature is as high as 900 °C – dwelling times of several hundreds of hours before equilibrium is reached. Reversibility has been shown, so this is not a phase change or decomposition phenomenon, but rather owed to rapidly decreasing oxygen exchange parameters D^{δ} and k^{δ} at these low oxygen partial pressures. These facts have to be considered when comparing literature values for MIEC materials in cases where no detailed information is given to the exact method and conditions of the measurement.

With the performed enhancements, the oxygen pump setup could be used to extract k^{δ} values for LSCF, PSCF and BSCF at 900 °C between 10^{-5} bar and 0.1 bar pO_2 not previously available in literature, and perfectly supplemented by the value for LSCF extracted at ambient air pressure by means of electrical impedance spectroscopy (EIS) measurements on an SOFC cathode. Parameter extraction on LSCF at 800 °C proved the setup and method to be accurate as shown by means of comparison with literature data. A visualization method comparing the mathematical fit to the measured sample response allows easy evaluation of the data quality. Ongoing work looking into adaptation of sample size, step change magnitude and in every case specific required controller parameter adaptation will enable measurements in an even larger pO_2 and temperature range as well as give results in closer proximity to the true equilibrium at a given partial pressure (one of the inherent problems with ECR techniques in the time domain).

Near-equilibrium values can be extracted by using ECR in the frequency domain. This was done on BSCF epitaxial and polycrystalline thin films as well as LSCF polycrystalline thin films in synthetic air at intermediate temperatures from 400 °C to 700 °C, further broadening the available field of information for these materials. The used mathematical model had formerly been developed for single-crystalline MIEC systems but has nonetheless been shown to be applicable for parameter extraction on these thin film samples, hence enabling near-equilibrium surface exchange parameter determination as opposed to the (by design) non-equilibrium measurements using ECR in the oxygen pump setup. Measurements on BSCF thin films revealed a higher activation energy exhibited by the polycrystalline film in comparison to the epitaxial ones, giving further evidence to the previous assumption of lessened substrate influence – and therefore better oxygen exchange at higher temperatures – on the polycrystalline system.

Furthermore, by performing ECR measurements in the frequency domain in the used kinetic set-up, changes in the thin film's surface performance due to the presence of selectable gaseous contaminants can be monitored in situ. For an epitaxial BSCF thin film, this was done concerning low concentrations of carbon dioxide in synthetic air: BSCF thin films degrade in atmospheres containing even small amounts of CO_2 (thrice the concentration of ambient air). Carbonate formation proved to be partly reversible; however, at higher amounts (already at 1 % CO_2) the thin film was irreversibly destroyed as revealed by combined SEM/EDX as well as TEM analyses.

Long term conductivity measurements on BSCF bulk samples (approx. 800 hours dwelling time at each temperature) revealed near-constant performance at 900 °C, while showing slow changes (degradation) in the BSCF system at 800 °C and a subsequent increase in conductivity after cooling to 700 °C. Significant changes in conductivity on the timescale of mere days during these long term measurements could be attributed to meteorological changes in ambient air pressure and somewhat corrected by application of knowledge pertaining to $\sigma(pO_2,T)$ gained through measurements in low oxygen partial pressures in the oxygen pump setup.

The detected decrease in electrical conductivity in these bulk samples triggered investigations into thermal phase stability by annealing BSCF powders in ambient air at different temperatures with subsequent XRD analyses. This revealed the occurrence of a hexagonal phase in the cubic BSCF system at 800 °C, leading to a phase equilibrium between cubic and hexagonal phase with a t_{90} of around 40 days. Annealing times of 72 hours revealed observable phase transition at temperatures below 840 °C that proved to be independent of any trace gases in the atmosphere. The hexagonal phase transforms back to purely cubic after subsequent annealing at 900 °C. Transition to the hexagonal phase also occurs at 700 °C, albeit with a different phase equilibrium, shifted more in favor of the cubic configuration in comparison to the equilibrium at 800 °C, explaining the detected increase in bulk conductivity measured in the long term experiment. Conductivity degradation is more pronounced in samples with smaller average grain size, indicating that phase transformation predominantly occurs at grain boundaries. This assumption has been verified by SEM/TEM analyses done at the Laboratory for Electron Microscopy (LEM), KIT, on annealed and quenched bulk samples.

In additional long term measurements, the degradation process at 800 °C could be seen to consist of two steps, the first being a faster degradation during the first 100 hours of the measurement, followed by the slow long term formation of the hexagonal phase.

SEM/TEM analyses at LEM showed this first, faster process to be the formation of plate-like regions, consisting of a lamellar substructure of cubic, hexagonal and layered barium cobalt oxide phases. Also revealed was the strong iron-depletion in the hexagonal phase, explaining the slow formation by the necessity of large-scale cation migration.

As all these secondary phases have detrimental influence on the possible oxygen permeation rate, efforts have been made to avoid phase transitions at temperatures ~800 °C. To this end, doping the B-site of the BSCF perovskite with small amounts of Y, Sc or Zr was used to stabilize the cubic matrix by introducing single-valent cations.

Annealing experiments and subsequent SEM/TEM analyses at LEM proved the BSCF3Y sample (3 at% Y-doping on the perovskite's B-site) to show no formation of the plate-like regions at all four temperatures (700, 800, 900 and 1000 °C) and therefore no decrease in electrical conductivity occurs during long term measurements in this temperature range. Stability improves even more by increasing the amount of doping to 10 at%. Literature data and preliminary permeation measurements suggest no negative impact on achievable oxygen flux through BSCFnY membranes for low dopant concentrations of up to 3 at%. Using this phase stabilization, membranes made from this material can be operated at temperatures lower than 900 °C; therefore increasing the efficiency in a membrane air separation reactor.

Yttrium dopant concentrations of 10 at%, however, result in a drop in oxygen permeation of approx. 20 %. This could be rectified by enhancing the effective surface exchange parameter k^{δ} of a membrane made from BSCF10Y, currently under investigation in the context of a DFG project[2]: investigations look into enhancing oxygen surface exchange rates of membranes by geometrically enlarging the active surface (enhancing the oxygen transfer through the surface by a decade) and/or introducing catalytically active surface modifications like hetero-interfaces of $(La,Sr)CoO_{3-\delta}$ / $(La,Sr)_2CoO_{4\pm\delta}$, capable of enhancing the oxygen transfer by a factor of 50[3].

Using this technique, the requested flux of 10 ml_N cm^{-2} min^{-1} (benchmark value formulated in the MEM-BRAIN project) should easily be possible and the introduction of high performance membrane modules for the Oxyfuel process in order to avoid further large-scale anthropogenic CO_2 emissions could become reality.

[2] Project “Leistungsfähige mischleitende Membranen mit nanoskaligen Funktionsschichten“ funded by the German Research Foundation (Deutsche Forschungsgemeinschaft, DFG) under grant no. IV 14/21-1.

[3] J. Hayd, H. Yokokawa, E. Ivers-Tiffée, "Hetero-Interfaces at Nanoscaled $(La,Sr)CoO_{3-\delta}$ Thin-Film Cathodes Enhancing Oxygen Surface-Exchange Properties", *J. Electrochem. Soc.* **160**, pp. F351-F359 (2013).

Appendix A

List of Samples

A.1 Powder Samples

Powder	Composition	Preparation Route	Prepared / Supplied by	Particle Size D_{50} / µm
BSCF	$Ba_{0.5}Sr_{0.5}Co_{0.8}Fe_{0.2}O_{3-\delta}$	Mixed Oxide	IKTS Hermsdorf	2.40
LSCF	$La_{0.58}Sr_{0.4}Co_{0.2}Fe_{0.8}O_{3-\delta}$	Spray Pyrolysis	IEK-1, FZ Jülich	2…3
LSC	$La_{0.6}Sr_{0.4}CoO_{3-\delta}$	Mixed Oxide	IWE	2…3
PSCF	$Pr_{0.58}Sr_{0.4}Co_{0.2}Fe_{0.8}O_{3-\delta}$	Spray Pyrolysis	IEK-1, FZ Jülich	2…3
BSCF3Y	$(Ba_{0.5}Sr_{0.5})(Co_{0.8}Fe_{0.2})_{0.97}Y_{0.03}O_{3-\delta}$	Mixed Oxide	IKTS Hermsdorf	7.1
BSCF3Sc	$(Ba_{0.5}Sr_{0.5})(Co_{0.8}Fe_{0.2})_{0.97}Sc_{0.03}O_{3-\delta}$	Mixed Oxide	IKTS Hermsdorf	6.3
BSCF3Zr	$(Ba_{0.5}Sr_{0.5})(Co_{0.8}Fe_{0.2})_{0.97}Zr_{0.03}O_{3-\delta}$	Mixed Oxide	IKTS Hermsdorf	5.8

A.2 Sintered Bulk Samples

Sintered bulk samples presented in this thesis are listed on the following pages. Please note that the tables are presented on a double-page spread.

Sample ID	Composition	Prepared / Supplied by	Pressing Conditions
FZJ-BSCF-02	$Ba_{0.5}Sr_{0.5}Co_{0.8}Fe_{0.2}O_{3-\delta}$	IEK1- FZJ (IKTS powder)	uniaxial with 10 kN/cm^2
FZJ-PSCF-02	$Pr_{0.58}Sr_{0.4}Co_{0.2}Fe_{0.8}O_{3-\delta}$	IEK1- FZJ (FZJ powder)	uniaxial with 10 kN/cm^2
FZJ-BSCF-05	$Ba_{0.5}Sr_{0.5}Co_{0.8}Fe_{0.2}O_{3-\delta}$	IEK1- FZJ (IKTS powder)	uniaxial with 10 kN/cm^2
B-BSCF-001-A	$Ba_{0.5}Sr_{0.5}Co_{0.8}Fe_{0.2}O_{3-\delta}$	IEK1- FZJ (IKTS powder)	uniaxial with 10 kN/cm^2
B-BSCF-001-B	$Ba_{0.5}Sr_{0.5}Co_{0.8}Fe_{0.2}O_{3-\delta}$	IEK1- FZJ (IKTS powder)	uniaxial with 10 kN/cm^2
B-BSCF-003-A	$Ba_{0.5}Sr_{0.5}Co_{0.8}Fe_{0.2}O_{3-\delta}$	IEK1- FZJ (IKTS powder)	uniaxial with 10 kN/cm^2
B-BSCF-003-B	$Ba_{0.5}Sr_{0.5}Co_{0.8}Fe_{0.2}O_{3-\delta}$	IEK1- FZJ (IKTS powder)	uniaxial with 10 kN/cm^2
B-BSCF-004-A	$Ba_{0.5}Sr_{0.5}Co_{0.8}Fe_{0.2}O_{3-\delta}$	IEK1- FZJ (IKTS powder)	uniaxial with 10 kN/cm^2
B-BSCF-004-B	$Ba_{0.5}Sr_{0.5}Co_{0.8}Fe_{0.2}O_{3-\delta}$	IEK1- FZJ (IKTS powder)	uniaxial with 10 kN/cm^2
B-LSCF-014	$La_{0.58}Sr_{0.4}Co_{0.2}Fe_{0.8}O_{3-\delta}$	IEK1- FZJ (FZJ powder)	uniaxial with 10 kN/cm^2
B-BSCF-026A	$Ba_{0.5}Sr_{0.5}Co_{0.8}Fe_{0.2}O_{3-\delta}$	IEK1- FZJ (IKTS powder)	uniaxial with 10 kN/cm^2
B-BSCF-028A	$Ba_{0.5}Sr_{0.5}Co_{0.8}Fe_{0.2}O_{3-\delta}$	IEK1- FZJ (IKTS powder)	uniaxial with 10 kN/cm^2
B-BSCF-028B	$Ba_{0.5}Sr_{0.5}Co_{0.8}Fe_{0.2}O_{3-\delta}$	IEK1- FZJ (IKTS powder)	uniaxial with 10 kN/cm^2
B-BSCF-029A	$Ba_{0.5}Sr_{0.5}Co_{0.8}Fe_{0.2}O_{3-\delta}$	IEK1- FZJ (IKTS powder)	uniaxial with 10 kN/cm^2
B-LSC-056	$La_{0.6}Sr_{0.4}CoO_{3-\delta}$	IWE (IWE powder)	isostatic with 20 kN/cm^2

Sintering Conditions (in Air)	**Density**	**Grain Size / µm**	**Sample Dimensions / mm³**	**Comments**
12 h at 1000 °C	> 95 %	~ 25	~ 10 x 6 x 1.2	cut by ultrasonic lapping, mechanically polished with P1200 emery paper, contacting: Pt wire & paste (1050 °C, 1 h)
5 h at 1250 °C	> 95 %	n/a	~ 10 x 6 x 1.1	contacted with Pt wire & paste (1050 °C, 1 h)
12 h at 1000 °C	> 95 %	~ 25	~ 10 x 6 x 1.2	cut by ultrasonic lapping, mechanically polished with P1200 emery paper, contacting: Pt wire & paste (1050 °C, 1 h)
12 h at 1000 °C	> 95 %	~ 25	~ 10 x 5 x 1	contacting: Pt wire & paste (1050 °C, 1 h)
12 h at 1000 °C	> 95 %	~ 25	~ 10 x 5 x 1	for structural analysis
50 h at 1150 °C	> 95 %	~ 90	~ 10 x 5 x 1	contacting: Pt wire & paste (1050 °C, 1 h)
50 h at 1150 °C	> 95 %	~ 90	~ 10 x 5 x 1	for structural analysis
50 h at 1120 °C	> 95 %	~ 50	~ 10 x 5 x 1	contacting: Pt wire & paste (1050 °C, 1 h)
50 h at 1120 °C	> 95 %	~ 50	~ 10 x 5 x 1	for structural analysis
5 h at 1200 °C	> 95 %	n/a	~ 10 x 6 x 1.1	contacted with Pt wire & paste (1050 °C, 1 h)
12 h at 1000 °C	> 95 %	~ 25	~ 5 x 5 x 1	quenched after 24 h at 1000 °C in air
12 h at 1000 °C	> 95 %	~ 25	~ 5 x 5 x 1	quenched after 24 h at 1000 °C and 100 h at 900 °C in air
12 h at 1000 °C	> 95 %	~ 25	~ 5 x 5 x 1	quenched after 24 h at 1000 °C and 100 h at 800 °C in air
12 h at 1000 °C	> 95 %	~ 25	~ 5 x 5 x 1	quenched after 24 h at 1000 °C and 100 h at 700 °C in air
12 h at 1350 °C	> 95 %	n/a	14.1 x 5.4 x 1.1	contacted with Pt wire & paste (1050 °C, 1 h)

Sample ID	Composition	Prepared / Supplied by	Pressing Conditions
B-BSCF3Zr-059	$(Ba_{0.5}Sr_{0.5})(Co_{0.8}Fe_{0.2})_{0.97}Zr_{0.03}O_{3-\delta}$	IWE (IWE powder)	isostatic with 27 kN/cm²
B-BSCF-3Y-119	$(Ba_{0.5}Sr_{0.5})(Co_{0.8}Fe_{0.2})_{0.97}Y_{0.03}O_{3-\delta}$	IWE (IKTS powder)	isostatic with 27 kN/cm²
B-BSCF-3Sc-121	$(Ba_{0.5}Sr_{0.5})(Co_{0.8}Fe_{0.2})_{0.97}Sc_{0.03}O_{3-\delta}$	IWE (IKTS powder)	isostatic with 27 kN/cm²
B-BSCF-3Zr-128	$(Ba_{0.5}Sr_{0.5})(Co_{0.8}Fe_{0.2})_{0.97}Zr_{0.03}O_{3-\delta}$	IWE (IKTS powder)	isostatic with 27 kN/cm²
B-BSCF-3Y-132	$(Ba_{0.5}Sr_{0.5})(Co_{0.8}Fe_{0.2})_{0.97}Y_{0.03}O_{3-\delta}$	IWE (IKTS powder)	isostatic with 27 kN/cm²
B-BSCF3Sc-133	$(Ba_{0.5}Sr_{0.5})(Co_{0.8}Fe_{0.2})_{0.97}Sc_{0.03}O_{3-\delta}$	IWE (IKTS powder)	isostatic with 27 kN/cm²
B-BSCF3Sc-134	$(Ba_{0.5}Sr_{0.5})(Co_{0.8}Fe_{0.2})_{0.97}Sc_{0.03}O_{3-\delta}$	IWE (IKTS powder)	isostatic with 27 kN/cm²
B-BSCF3Zr-135	$(Ba_{0.5}Sr_{0.5})(Co_{0.8}Fe_{0.2})_{0.97}Zr_{0.03}O_{3-\delta}$	IWE (IKTS powder)	isostatic with 27 kN/cm²
B-BSCF3Zr-136	$(Ba_{0.5}Sr_{0.5})(Co_{0.8}Fe_{0.2})_{0.97}Zr_{0.03}O_{3-\delta}$	IWE (IKTS powder)	isostatic with 27 kN/cm²
B-BSCF3Y-139	$(Ba_{0.5}Sr_{0.5})(Co_{0.8}Fe_{0.2})_{0.97}Y_{0.03}O_{3-\delta}$	IWE (IKTS powder)	isostatic with 27 kN/cm²
B-BSCF3Y-140	$(Ba_{0.5}Sr_{0.5})(Co_{0.8}Fe_{0.2})_{0.97}Y_{0.03}O_{3-\delta}$	IWE (IKTS powder)	isostatic with 27 kN/cm²
B-BSCF3Sc-141	$(Ba_{0.5}Sr_{0.5})(Co_{0.8}Fe_{0.2})_{0.97}Sc_{0.03}O_{3-\delta}$	IWE (IKTS powder)	isostatic with 27 kN/cm²
B-BSCF3Sc-142	$(Ba_{0.5}Sr_{0.5})(Co_{0.8}Fe_{0.2})_{0.97}Sc_{0.03}O_{3-\delta}$	IWE (IKTS powder)	isostatic with 27 kN/cm²
B-BSCF3Zr-143	$(Ba_{0.5}Sr_{0.5})(Co_{0.8}Fe_{0.2})_{0.97}Zr_{0.03}O_{3-\delta}$	IWE (IKTS powder)	isostatic with 27 kN/cm²
B-BSCF3Zr-144	$(Ba_{0.5}Sr_{0.5})(Co_{0.8}Fe_{0.2})_{0.97}Zr_{0.03}O_{3-\delta}$	IWE (IKTS powder)	isostatic with 27 kN/cm²
B-BSCF-145	$Ba_{0.5}Sr_{0.5}Co_{0.8}Fe_{0.2}O_{3-\delta}$	IEK1- FZJ (IKTS powder)	uniaxial with 10 kN/cm²

Sintering Conditions (in Air)	**Density**	**Grain Size / µm**	**Sample Dimensions / mm³**	**Comments**
12 h at 1050 °C	n/a	n/a	~ 15 x 6 x 1.6	contacted by DC sputtering with Pt and Pt wire & paste (1050 °C, 1 h)
12 h at 1050 °C	n/a	n/a	n/a	for structural analysis
12 h at 1050 °C	n/a	n/a	n/a	for structural analysis
12 h at 1050 °C	n/a	n/a	n/a	for structural analysis
12 h at 1100 °C	n/a	n/a	n/a	for structural analysis
12 h at 1100 °C	89 %	n/a	~ 15 x 6 x 1.6	contacted by DC sputtering with Pt and Pt wire & paste (1050 °C, 1 h)
12 h at 1100 °C	89 %	n/a	~ 15 x 6 x 1.6	contacted by DC sputtering with Pt and Pt wire & paste (1050 °C, 1 h)
12 h at 1100 °C	85 %	n/a	~ 15 x 6 x 1.6	contacted by DC sputtering with Pt and Pt wire & paste (1050 °C, 1 h)
12 h at 1100 °C	85 %	n/a	n/a	for structural analysis
12 h at 1150 °C	86 %	n/a	~ 15 x 6 x 1.7	contacted by DC sputtering with Pt and Pt wire & paste (1050 °C, 1 h)
12 h at 1150 °C	86 %	n/a	~ 15 x 6 x 1.3	contacted by DC sputtering with Pt and Pt wire & paste (1050 °C, 1 h)
12 h at 1150 °C	n/a	n/a	~ 15 x 6 x 1.8	for structural analysis
12 h at 1150 °C	n/a	n/a	~ 15 x 6 x 0.6	for structural analysis
12 h at 1150 °C	n/a	n/a	~ 15 x 6 x 1.6	for structural analysis
12 h at 1150 °C	n/a	n/a	~ 15 x 6 x 0.9	for structural analysis
12 h at 1000 °C	> 95 %	~ 25	~ 12 x 6 x 2	contacted by DC sputtering with Pt and Pt wire & paste (1050 °C, 1 h)

Sample ID	Composition	Prepared / Supplied by	Pressing Conditions
B-BSCF-146	$Ba_{0.5}Sr_{0.5}Co_{0.8}Fe_{0.2}O_{3-\delta}$	IEK1- FZJ (IKTS powder)	uniaxial with 10 kN/cm²
B-BSCF-147	$Ba_{0.5}Sr_{0.5}Co_{0.8}Fe_{0.2}O_{3-\delta}$	IEK1- FZJ (IKTS powder)	uniaxial with 10 kN/cm²
B-BSCF-148	$Ba_{0.5}Sr_{0.5}Co_{0.8}Fe_{0.2}O_{3-\delta}$	IEK1- FZJ (IKTS powder)	uniaxial with 10 kN/cm²
B-BSCF3Zr-151A	$(Ba_{0.5}Sr_{0.5})(Co_{0.8}Fe_{0.2})_{0.97}Zr_{0.03}O_{3-\delta}$	IWE (IKTS powder)	isostatic with 27 kN/cm²
B-BSCF3Sc-155B	$(Ba_{0.5}Sr_{0.5})(Co_{0.8}Fe_{0.2})_{0.97}Sc_{0.03}O_{3-\delta}$	IWE (IKTS powder)	isostatic with 27 kN/cm²
B-BSCF3Sc-156A	$(Ba_{0.5}Sr_{0.5})(Co_{0.8}Fe_{0.2})_{0.97}Sc_{0.03}O_{3-\delta}$	IWE (IKTS powder)	isostatic with 27 kN/cm²
B-BSCF3Sc-157A	$(Ba_{0.5}Sr_{0.5})(Co_{0.8}Fe_{0.2})_{0.97}Sc_{0.03}O_{3-\delta}$	IWE (IKTS powder)	isostatic with 27 kN/cm²
B-BSCF-3Y-161B	$(Ba_{0.5}Sr_{0.5})(Co_{0.8}Fe_{0.2})_{0.97}Y_{0.03}O_{3-\delta}$	IWE (IKTS powder)	isostatic with 27 kN/cm²
B-BSCF-3Y-162A	$(Ba_{0.5}Sr_{0.5})(Co_{0.8}Fe_{0.2})_{0.97}Y_{0.03}O_{3-\delta}$	IWE (IKTS powder)	isostatic with 27 kN/cm²
B-BSCF3Y-163A	$(Ba_{0.5}Sr_{0.5})(Co_{0.8}Fe_{0.2})_{0.97}Y_{0.03}O_{3-\delta}$	IWE (IKTS powder)	isostatic with 27 kN/cm²
B-BSCF-3Zr-205 B	$(Ba_{0.5}Sr_{0.5})(Co_{0.8}Fe_{0.2})_{0.97}Zr_{0.03}O_{3-\delta}$	IWE (IKTS powder)	isostatic with 27 kN/cm²
B-BSCF-3Zr-207 A	$(Ba_{0.5}Sr_{0.5})(Co_{0.8}Fe_{0.2})_{0.97}Zr_{0.03}O_{3-\delta}$	IWE (IKTS powder)	isostatic with 27 kN/cm²
B-BSCF-225	$Ba_{0.5}Sr_{0.5}Co_{0.8}Fe_{0.2}O_{3-\delta}$	IWE (IWE powder)	isostatic with 27 kN/cm²
B-BSCF-3Y-242	$(Ba_{0.5}Sr_{0.5})(Co_{0.8}Fe_{0.2})_{0.97}Y_{0.03}O_{3-\delta}$	IWE (IWE powder)	isostatic with 27 kN/cm²
B-BSCF-10Y-246	$(Ba_{0.5}Sr_{0.5})(Co_{0.8}Fe_{0.2})_{0.9}Y_{0.1}O_{3-\delta}$	IWE (IWE powder)	isostatic with 27 kN/cm²
B-BSCF-249	$Ba_{0.5}Sr_{0.5}Co_{0.8}Fe_{0.2}O_{3-\delta}$	IWE (IWE powder)	isostatic with 27 kN/cm²

Sintering Conditions (in Air)	Density	Grain Size / µm	Sample Dimensions / mm³	Comments
12 h at 1000 °C	> 95 %	~ 25	~ 12 x 6 x 2	contacted by DC sputtering with Pt and Pt wire & paste (1050 °C, 1 h)
12 h at 1000 °C	> 95 %	~ 25	~ 12 x 6 x 2	contacted by DC sputtering with Pt and Pt wire & paste (1050 °C, 1 h)
12 h at 1000 °C	> 95 %	~ 25	~ 12 x 6 x 2	contacted by DC sputtering with Pt and Pt wire & paste (1050 °C, 1 h)
12 h at 1100 °C	n/a	n/a	~ 5 x 5 x 1	quenched after 10 h at 1000 °C and 100 h at 700 °C in air
12 h at 1100 °C	n/a	n/a	~ 5 x 5 x 1	quenched after 10 h at 1000 °C and 100 h at 900 °C in air
12 h at 1100 °C	n/a	n/a	~ 5 x 5 x 1	quenched after 10 h at 1000 °C and 100 h at 800 °C in air
12 h at 1100 °C	n/a	n/a	~ 5 x 5 x 1	quenched after 10 h at 1000 °C and 100 h at 700 °C in air
12 h at 1100 °C	n/a	n/a	~ 5 x 5 x 1	quenched after 10 h at 1000 °C and 100 h at 900 °C in air
12 h at 1100 °C	n/a	n/a	~ 5 x 5 x 1	quenched after 10 h at 1000 °C and 100 h at 800 °C in air
12 h at 1100 °C	n/a	n/a	~ 5 x 5 x 1	quenched after 10 h at 1000 °C and 100 h at 700 °C in air
12 h at 1100 °C	n/a	n/a	~ 5 x 5 x 1	quenched after 24 h at 1000 °C and 100 h at 900 °C in air
12 h at 1100 °C	n/a	n/a	~ 5 x 5 x 1	quenched after 24 h at 1000 °C and 100 h at 800 °C in air
12 h at 1050 °C	n/a	n/a	~ 15 x 6 x 2	contacted with Pt wire & paste (1050 °C, 1 h) for long term meas.
12 h at 1150 °C	n/a	n/a	~ 15 x 6 x 2	contacted with Pt wire & paste (1050 °C, 1 h) for long term meas.
12 h at 1150 °C	n/a	n/a	~ 15 x 6 x 2	contacted with Pt wire & paste (1050 °C, 1 h) for long term meas.
12 h at 1050 °C	n/a	n/a	~ 15 x 6 x 2	contacted with Pt wire & paste (1050 °C, 1 h) for long term meas.

A.3 Thin Films on NGO (110) Single-Crystalline Substrates

Sintered thin film samples presented in this thesis are listed on the following pages. Please note that the table is presented on a double-page spread.

Sample ID	Composition	Target Preparation
PLD-BSCF-11a	$Ba_{0.5}Sr_{0.5}Co_{0.8}Fe_{0.2}O_{3-\delta}$	IWE (sintered at 1100 °C, 10 h, from IWE-BSCF-01-3100h powder)
PLD-BSCF-17n1	$Ba_{0.5}Sr_{0.5}Co_{0.8}Fe_{0.2}O_{3-\delta}$	IWE (sintered at 1100 °C, 10 h, from IWE-BSCF-01-3100h powder)
PLD-BSCF-17n2	$Ba_{0.5}Sr_{0.5}Co_{0.8}Fe_{0.2}O_{3-\delta}$	IWE (sintered at 1100 °C, 10 h, from IWE-BSCF-01-3100h powder)
PLD-BSCF-18n1	$Ba_{0.5}Sr_{0.5}Co_{0.8}Fe_{0.2}O_{3-\delta}$	IWE (sintered at 1100 °C, 10 h, from IWE-BSCF-01-3100h powder)
PLD-BSCF-18n2	$Ba_{0.5}Sr_{0.5}Co_{0.8}Fe_{0.2}O_{3-\delta}$	IWE (sintered at 1100 °C, 10 h, from IWE-BSCF-01-3100h powder)
PLD-BSCF-19n1	$Ba_{0.5}Sr_{0.5}Co_{0.8}Fe_{0.2}O_{3-\delta}$	IWE (sintered at 1100 °C, 10 h, from IWE-BSCF-01-3100h powder)
PLD-BSCF-19n2	$Ba_{0.5}Sr_{0.5}Co_{0.8}Fe_{0.2}O_{3-\delta}$	IWE (sintered at 1100 °C, 10 h, from IWE-BSCF-01-3100h powder)
PLD-BSCF-23n2	$Ba_{0.5}Sr_{0.5}Co_{0.8}Fe_{0.2}O_{3-\delta}$	IWE (sintered at 1100 °C, 10 h, from IWE-BSCF-01-3100h powder)
MOD-BSCF-066	$Ba_{0.5}Sr_{0.5}Co_{0.8}Fe_{0.2}O_{3-\delta}$	from MOD solution (Kojundo, Japan)
MOD-BSCF-068	$Ba_{0.5}Sr_{0.5}Co_{0.8}Fe_{0.2}O_{3-\delta}$	from MOD solution (Kojundo, Japan)
D-LSCF-028	$La_{0.58}Sr_{0.4}Co_{0.2}Fe_{0.8}O_{3-\delta}$	from MOD solution (ISC, Würzburg)
D-LSCF-036	$La_{0.58}Sr_{0.4}Co_{0.2}Fe_{0.8}O_{3-\delta}$	powder target (IEK-1, FZJ)
D-LSCF-037	$La_{0.58}Sr_{0.4}Co_{0.2}Fe_{0.8}O_{3-\delta}$	powder target (IEK-1, FZJ)

Preparation (ISC, IEK-1, CINC2, IWE)	**Film Thickness / nm**	**Comments**
CINC2, at 700 °C in 0.4 mbar O_2, 10000 laser pulses	340	Pt-contacted (2-wire) (1050 °C, 1 h)
CINC2, at 700 °C in 0.4 mbar O_2, 10000 laser pulses	340	Pt-contacted (2-wire) (1050 °C, 1 h)
CINC2, at 700 °C in 0.4 mbar O_2, 10000 laser pulses	340	Pt-contacted (2-wire) (1050 °C, 1 h)
CINC2, at 700 °C in 0.4 mbar O_2, 10000 laser pulses	340	Au-contacted (2-wire) (900 °C, 1 h)
CINC2, at 700 °C in 0.4 mbar O_2, 10000 laser pulses	340	Au-contacted (2-wire) (900 °C, 1 h)
CINC2, at 700 °C in 0.4 mbar O_2, 4000 laser pulses	136	not contacted
CINC2, at 700 °C in 0.4 mbar O_2, 4000 laser pulses	136	not contacted
CINC2, at 700 °C in 0.4 mbar O_2, 10000 laser pulses	340	Au-contacted (2-wire) (900 °C, 1 h)
IWE (Dr. Asano), 5-time spin-coated, annealed at 950 °C	~ 50	Au-contacted by DC sputtering
IWE (Dr. Asano), 5-time spin-coated, annealed at 950 °C	~ 50	Au-contacted by DC sputtering
ISC, spin-coated, annealed 1 h at 1000 °C in air	125	Pt-contacted (2-wire) (1050 °C, 1 h)
IEK-1, Sputtering	~ 1000	Pt-contacted (2-wire) (1050 °C, 1 h)
IEK-1, Sputtering	~ 1000	Pt-contacted (2-wire) (1050 °C, 1 h)

Appendix B

Publications

The results obtained in this thesis have been presented to representatives from science and industry by publications in international journals and presentations at national and international conferences. This section contains a list of those publications and conference presentations.

B.1 Journal Papers and Conference Proceedings

- C. Niedrig; S. F. Wagner; W. Menesklou; S. Baumann; and E. Ivers-Tiffée, "Oxygen Equilibration Kinetics of Mixed-Conducting Perovskites BSCF, LSCF, and PSCF at 900 °C", *Solid State Ionics*, in press (2015).
- C. Endler-Schuck, J. Joos, C. Niedrig, A. Weber, and E. Ivers-Tiffée, "The Chemical Oxygen Surface Exchange and Bulk Diffusion Coefficient Determined by Impedance Spectroscopy of Porous $La_{0.58}Sr_{0.4}Co_{0.2}Fe_{0.8}O_{3-\delta}$ (LSCF) Cathodes", *Solid State Ionics* **269** (2015), 67-79.
- C. Niedrig, S. F. Wagner, W. Menesklou, and E. Ivers-Tiffée, "Characterization of Oxygen-Dependent Stability of Selected Mixed-Conducting Perovskite Oxides", *Solid State Ionics* **273** (2015), 41-45.
- K. Asano, C. Niedrig, W. Menesklou, S. F. Wagner and E. Ivers-Tiffée, "Modification of Oxygen / $(Ba_{0.5}Sr_{0.5})(Co_{0.8}Fe_{0.2})O_{3-\delta}$ Interfaces Derived By Metal-Organic Deposition", *ECS Transactions* **66 (2015)**, 147-155.

- A. Häffelin, C. Niedrig, S. F. Wagner, S. Baumann, W. A. Meulenberg, and E. Ivers-Tiffée, "Three-Dimensional Performance Model for Oxygen Transport Membranes", *J. Electrochem. Soc.* **161** (2014), F1409-F1415.
- E. Ivers-Tiffée, C. Niedrig, and S. F. Wagner, "MIEC Materials for Membrane Applications: Enhancing the Oxygen Transport", *ECS Trans.* **61** (2014), 283-293.
- A. Häffelin, C. Niedrig, S. F. Wagner, and E. Ivers-Tiffée, "Three-Dimensional Performance Model for Oxygen Transport Membranes", *ECS Trans.* **57** (2013), 2543-2552.
- P. Müller, H. Störmer, M. Meffert, L. Dieterle, C. Niedrig, S. F. Wagner, E. Ivers-Tiffée, and D. Gerthsen, "Secondary phase formation in $Ba_{0.5}Sr_{0.5}Co_{0.8}Fe_{0.2}O_{3-d}$ studied by electron microscopy", *Chem. Mater.* **25** (2013), 564-573.
- C. Niedrig, W. Menesklou, S. F. Wagner, and E. Ivers-Tiffée, "High-Temperature pO_2 Stability of Metal Oxides Determined by Amperometric Oxygen Titration", *J. Electrochem. Soc.* **160** (2013), F135-F140.
- P. Müller, H. Störmer, L. Dieterle, C. Niedrig, E. Ivers-Tiffée, and D. Gerthsen, "Decomposition pathway of cubic $Ba_{0.5}Sr_{0.5}Co_{0.8}Fe_{0.2}O_{3-d}$ between 700 °C and 1000 °C analyzed by electron microscopic techniques", *Solid State Ionics* **206** (2012), 57-66.
- C. Niedrig, P. Braun, W. Menesklou, S. F. Wagner, and E. Ivers-Tiffée, "Investigation of MIEC Materials using an Amperometric Oxygen Titration Method", *ECS Trans.* **45** (2012), 75-85.
- C. Niedrig, S. Taufall, M. Burriel, W. Menesklou, S. F. Wagner, S. Baumann, and E. Ivers-Tiffée, "Thermal stability of the cubic phase in $Ba_{0.5}Sr_{0.5}Co_{0.8}Fe_{0.2}O_{3-\delta}$ (BSCF)", *Solid State Ionics* **197** (2011), 25-31.
- S. F. Wagner, S. Taufall, C. Niedrig, H. Götz, W. Menesklou, S. Baumann, and E. Ivers-Tiffée, "pO_2 stability of $Ba_{0.5}Sr_{0.5}Co_{0.8}Fe_{0.2}O_{3-\delta}$", *MRS Symp. Proc. Vol.* **1309** (2011), 57-64.
- M. Burriel, C. Niedrig, W. Menesklou, S. F. Wagner, J. Santiso, and E. Ivers-Tiffée, "BSCF epitaxial thin films: Electrical transport and oxygen surface exchange", *Solid State Ionics* **181** (2010), 602-608.
- P. Müller, L. Dieterle, E. Müller, H. Störmer, D. Gerthsen, C. Niedrig, S. Taufall, S. F. Wagner, and E. Ivers-Tiffée, "$Ba_{0.5}Sr_{0.5}Co_{0.8}Fe_{0.2}O_{3-d}$ for Oxygen Separation Membranes", *ECS Trans.* **28** (2010), 309-314.
- C. Niedrig, S. Taufall, S. F. Wagner, P. Müller, H. Störmer, D. Gerthsen, and E. Ivers-Tiffée, "Thermal Stability of mixed ionic-electronic conducting $Ba_{0.5}Sr_{0.5}Co_{0.8}Fe_{0.2}O_{3-\delta}$", in Proc. of the 1st International Conference on Materials for Energy, Karlsruhe/Germany, Book A: Dechema (2010), 48-50.

- P. Müller, H. Störmer, D. Gerthsen, J. Santiso, J. Roqueta, M. Burriel, C. Niedrig, S. F. Wagner, and E. Ivers-Tiffée, "TEM investigations of mixed ionic/electronic conducting $Ba_{0.5}Sr_{0.5}Co_{0.8}Fe_{0.2}O_{3-\delta}$ materials", in W. Grogger, F. Hofer, and P. Pölt (Eds.), Proc. MC2009 **3**, Graz, Austria: Verlag der TU Graz (2009), 295-296.

B.2 Conference Presentations and Posters

- K. Asano, C. Niedrig, W. Menesklou, S. F. Wagner, and E. Ivers-Tiffée, "Modification of Oxygen/$(Ba_{0.5}Sr_{0.5})(Co_{0.8}Fe_{0.2})O_{3-\delta}$ Interfaces Derived By Metal-Organic Deposition", abstract submitted for oral presentation at the *227th Meeting of The Electrochemical Society (ECS),* Chicago IL/USA, May 2015.
- S. F. Wagner, C. Niedrig, and E. Ivers-Tiffée, "Electrochemical Characterization and Performance Modeling of MIEC Materials for Oxygen-Transport Membranes", oral presentation at the *Materials Science Engineering (MSE 2014)* conference, Darmstadt/Germany, September 2014.
- M. Meffert, P. Müller, H. Störmer, L.-S. Unger, C. Niedrig, S. F. Wagner, S. Saher, H. Bouwmeester, and E. Ivers-Tiffée, D. Gerthsen, "Effect of yttrium (Y) and zirconium (Zr) doping on the thermodynamical stability of the cubic $Ba_{0.5}Sr_{0.5}Co_{0.8}Fe_{0.2}O_{3-\delta}$ phase", poster presentation at the *Microscopy & Microanalysis 2014* conference, Hartford CT/USA, August 2014.
- L.-S. Unger, C. Niedrig, W. Menesklou, S. F. Wagner, and E. Ivers-Tiffée, "Stability of Selected Mixed Conducting Perovskite Oxides at Low Oxygen Partial Pressures", poster presentation at the*13th International Conference on Inorganic Membranes (ICIM 2014)*, Brisbane/Australia, July 2014.
- M. Meffert, P. Müller, H. Störmer, L.-S. Unger, C. Niedrig, S. F. Wagner, E. Ivers-Tiffée, and D. Gerthsen, "Effect of Yttrium (Y) doping on the thermodynamical stability of the cubic $Ba_{0.5}Sr_{0.5}Co_{0.8}Fe_{0.2}O_{3-\delta}$ phase", poster presentation at the *13th International Conference on Inorganic Membranes (ICIM 2014)*, Brisbane/Australia, July 2014.
- L.-S. Unger, M. Meffert, H. Störmer, C. Niedrig, S. F. Wagner, W. Menesklou, D. Gerthsen, and E. Ivers-Tiffée, "Surface activation of $(Ba_{0.5}Sr_{0.5})(Co_{0.8}Fe_{0.2})O_{3-\delta}$ by means of a $(La_{0.6}Sr_{0.4})CoO_{3-\delta}$ functional layer", oral presentation at the *Electroceramics XIV Conference*, Bucharest/Romania, June 2014.

- M. Meffert, P. Müller, H. Störmer, L.-S. Unger, C. Niedrig, S. F. Wagner, E. Ivers-Tiffée, and D. Gerthsen, "Effect of yttrium doping on the cubic phase of $Ba_{0.5}Sr_{0.5}Co_{0.8}Fe_{0.2}O_{3-\delta}$ studied by electron microscopy", oral presentation at the *Electroceramics XIV Conference*, Bucharest/Romania, June 2014.
- L.-S. Unger, C. Niedrig, W. Menesklou, S. F. Wagner, and E. Ivers-Tiffée, "pO_2 Stability of Mixed Conducting Perovskites", poster presentation at the *Electroceramics XIV Conference*, Bucharest/Romania, June 2014.
- L.-S. Unger, M. Meffert, C. Niedrig, H. Störmer, W. Menesklou, S. F. Wagner, D. Gerthsen, and E. Ivers-Tiffée, "Enhanced Oxygen Surface Reaction in $(Ba_{0.5}Sr_{0.5})(Co_{0.8}Fe_{0.2})O_{3-\delta}$ by Nanoscaled $(La_{0.6}Sr_{0.4})CoO_{3-\delta}$ Functional Layer", oral presentation at the *E-MRS Spring Meeting 2014*, Lille/France, May 2014.
- C. Niedrig, L. Wolff, L.-S. Unger, W. Menesklou, S. F. Wagner, and E. Ivers-Tiffée, "High-Temperature Setup for Hall Measurements on Mixed Conductors", poster presentation at the *E-MRS Spring Meeting 2014*, Lille/France, May 2014.
- C. Niedrig, L. Willms, L.-S. Unger, W. Menesklou, S. F. Wagner, and E. Ivers-Tiffée, "Evaluation of ECR Measurements on MIEC Oxides Performed in an 'Oxygen Pump' Setup", poster presentation at the *E-MRS Spring Meeting 2014*, Lille/France, May 2014.
- C. Niedrig, L.-S. Unger, S. F. Wagner, W. Menesklou, and E. Ivers-Tiffée, "Characterization of Oxygen Stoichiometry Changes of Selected Mixed Conducting Perovskites", poster presentation at the *E-MRS Spring Meeting 2014*, Lille/France, May 2014.
- E. Ivers-Tiffée, C. Niedrig, and S. F. Wagner, "MIEC Materials for Membrane Applications: Enhancing the Oxygen Transport", invited talk given at the *IMCC 9 Symposium* of the *225th Meeting of The Electrochemical Society (ECS)*, Orlando FL/USA, May 2014.
- W. Menesklou, A. Häffelin, C. Niedrig, S. F. Wagner, and E. Ivers-Tiffée, "Electrochemical Characterization and Modeling of Oxygen-Transport Membranes based on MIEC Materials", invited talk given at *THERMEC' 2013 - International Conference on Processing & Manufacturing of Advanced Materials*, Las Vegas NV/USA, December 2013.
- L. Unger, C. Niedrig, S. F. Wagner, W. Menesklou, and E. Ivers-Tiffée, "Sintering behavior and electrical characterization of $(Ba_{0.5}Sr_{0.5})(Co_{0.8}Fe_{0.2})_{1-y}X_yO_{3-\delta}$ (X = Y, Zr, y = 0.01…0.1)", poster presentation at the *MRS Fall Meeting 2013*, Boston MA/USA, December 2013.

- A. Häffelin, C. Niedrig, S. Wagner, A. Weber, and E. Ivers-Tiffée, "Three-Dimensional Performance Model for Oxygen Transport Membranes", oral presentation at the *SOFC-XIII Conference*, Okinawa/Japan, October 2013.
- M. Meffert, P. Müller, H. Störmer, C. Niedrig, S. F. Wagner, E. Ivers-Tiffée, and D. Gerthsen, "Influence of yttrium doping on the stability of the cubic $Ba_{0.5}Sr_{0.5}Co_{0.8}Fe_{0.2}O_{3-\delta}$ phase studied by electron microscopy", poster presentation at the *Summer School on "Inorganic membranes for green chemical production and clean power generation"*, Valencia/Spain, September 2013.
- S. F. Wagner, A. Häffelin, C. Niedrig, W. Menesklou, and E. Ivers-Tiffée, "MIEC Materials for Membrane Applications: Electrochemical Characterization and Performance Modeling", invited talk at the *Summer School on "Inorganic membranes for green chemical production and clean power generation"*, Valencia/Spain, September 2013.
- M. Meffert, P. Müller, H. Störmer, C. Niedrig, S. F. Wagner, E. Ivers-Tiffée, and D. Gerthsen, "Electron microscopy study of Yttrium-doped $Ba_{0.5}Sr_{0.5}Co_{0.8}Fe_{0.2}O_{3-\delta}$", poster presentation at the *Microscopy Conference (MC-2013)*, Regensburg/Germany, August 2013.
- A. Häffelin, C. Niedrig, S. F. Wagner, and E. Ivers-Tiffée, "3D performance model for oxygen transport membranes", oral presentation at the *19th International Conference on Solid State Ionics (SSI-19)*, Kyoto/Japan, June 2013.
- C. Niedrig, S. F. Wagner, W. Menesklou, and E. Ivers-Tiffée, "Stability of Selected MIEC Perovskites at Low Oxygen Partial Pressures", poster presentation at the *19th International Conference on Solid State Ionics (SSI-19)*, Kyoto/Japan, June 2013.
- C. Niedrig, W. Menesklou, S. F. Wagner, C.-Y. Yoo, H. J. M. Bouwmeester, and E. Ivers-Tiffée, "Oxygen Transport Parameters of co-doped $(Ba_{0.5}Sr_{0.5})(Co_{0.8}Fe_{0.2})_{0.97}X_{0.03}O_{3-\delta}$ (X = Y, Zr)", poster presentation at the *19th International Conference on Solid State Ionics (SSI-19)*, Kyoto/Japan, June 2013.
- M. Meffert, P. Müller, H. Störmer, C. Niedrig, S. F. Wagner, E. Ivers-Tiffée, and D. Gerthsen, "Electron Microscopy Study of Yttrium-doped $(Ba_{0.5}Sr_{0.5})(Co_{0.8}Fe_{0.2})_{0.97}Y_{0.03}O_{3-\delta}$", oral presentation at the *19th International Conference on Solid State Ionics (SSI-19)*, Kyoto/Japan, June 2013.
- K. Asano, C. Niedrig, S. F. Wagner, W. Menesklou, and E. Ivers-Tiffée, "Characterization of Nanoscale Surface Modifications of $(Ba_{0.5}Sr_{0.5})(Co_{0.8}Fe_{0.2})O_{3-\delta}$ Thin Films Made by a MOD Method", poster presentation at the *19th International Conference on Solid State Ionics (SSI-19)*, Kyoto/Japan, June 2013.

- C. Niedrig, W. Menesklou, S. F. Wagner, S. Baumann, and E. Ivers-Tiffée, "Oxygen Stoichiometry Changes and pO_2-dependent Stability of Selected MIEC Perovskites", oral presentation at the *2nd International Conference on Materials for Energy (EnMat-II)*, Karlsruhe/Germany, May 2013.
- A. Häffelin, C. Niedrig, J. Joos, S. Wagner, and E. Ivers-Tiffée, "3D performance model for mixed conducting membranes", poster presentation at the *2nd International Conference on Materials for Energy (EnMat-II)*, Karlsruhe/Germany, May 2013.
- M. Meffert, P. Müller, H. Störmer, C. Niedrig, S. F. Wagner, E. Ivers-Tiffée, and D. Gerthsen, "Electron microscopy study of Yttrium-doped BSCF", oral presentation at the *2nd International Conference on Materials for Energy (EnMat-II)*, Karlsruhe/Germany, May 2013.
- C. Niedrig, W. Menesklou, S. F. Wagner, and E. Ivers-Tiffée, "An Amperometric Oxygen Titration Method for pO_2 Stability Investigations on High-Temperature MIEC Oxides", poster presentation at the *Bunsentagung 2013*, Karlsruhe/Germany, May 2013.
- M. Meffert, P. Müller, H. Störmer, C. Niedrig, S. F. Wagner, E. Ivers-Tiffée, and D. Gerthsen, "Electron microscopy study of Y-doped BSCF", oral presentation at the *DPG-Frühjahrstagung*, Regensburg/Germany, March 2013.
- P. Müller, B. Albrecht, H. Störmer, D. Fuchs, C. Niedrig, S. F. Wagner, and E. Ivers-Tiffée, D. Gerthsen, "Electron microscopy study of surface-functionalized $Ba_{0.5}Sr_{0.5}Co_{0.8}Fe_{0.2}O_{3-\delta}$ membranes for oxygen separation", poster presentation at the *15th European Microscopy Congress*, Manchester/UK, September 2012.
- M. Meffert, P. Müller, H. Störmer, C. Niedrig, S. F. Wagner, and E. Ivers-Tiffée, D. Gerthsen, "Electron microscopy study of $(Ba_{0.5}Sr_{0.5})(Co_{0.8}Fe_{0.2})_{0.97}X_{0.03}O_{3-\delta}$ (X = Zr, Y, Sc)", poster presentation at the *15th European Microscopy Congress*, Manchester/UK, September 2012.
- P. Müller, M. Meffert, H. Störmer, C. Niedrig, S. F. Wagner, E. Ivers-Tiffée, and D. Gerthsen, "Analysis of co-doped $(Ba_{0.5}Sr_{0.5})(Co_{0.8}Fe_{0.2})_{0.97}X_{0.03}O_{3-\delta}$ (BSCFX5582, X=Zr, Y, Sc) by Electron Microscopy", poster presentation at the *International Conference on Inorganic Membranes (ICIM-2012)*, Enschede/The Netherlands, July 2012.
- W. Menesklou, C. Niedrig, S. F. Wagner, and E. Ivers-Tiffée, "Stability and Conductivity vs. Oxygen Partial Pressure of Mixed Ionic-Electronic Conductors", oral presentation at *Electronic Materials Conference (EMC)*, Penn State University, June 2012.
- C. Niedrig, C.-Y. Yoo, W. Menesklou, S. Baumann, S. F. Wagner, H. J. M. Bouwmeester, and E. Ivers-Tiffée, "Equilibration Kinetics of Mixed Ionic-Electronic Conducting Materials at Different Oxygen Partial Pressures", oral presentation at *Electroceramics-XIII*, Enschede/The Netherlands, June 2012.

- P. Müller, M. Meffert, H. Störmer, C. Niedrig, S. F. Wagner, E. Ivers-Tiffée, and D. Gerthsen, "Electron Microscopy investigations of zirconium-, yttrium- and scandium-doped $(Ba_{0.5}Sr_{0.5})(Co_{0.8}Fe_{0.2})_{0.97}X_{0.03}O_{3-\delta}$ (BSCFX5582)", oral presentation at *Electroceramics-XIII*, Enschede/The Netherlands, June 2012.
- C. Niedrig, P. Müller, W. Menesklou, S. F. Wagner, H. Störmer, D. Gerthsen, and E. Ivers-Tiffée, "Enhanced Oxygen Surface Exchange in $Ba_{0.5}Sr_{0.5}Co_{0.8}Fe_{0.2}O_{3-\delta}$", poster presentation at *Electroceramics-XIII*, Enschede/The Netherlands, June 2012.
- C. Niedrig, W. Menesklou, S. F. Wagner, and E. Ivers-Tiffée, "Stability of Mixed Conducting Perovskites in Reducing Atmospheres", poster presentation at the *E-MRS Spring Meeting 2012*, Strasbourg/France, May 2012.
- C. Niedrig, P. Braun, W. Menesklou, S. F. Wagner, and E. Ivers-Tiffée, "Coulometric Oxygen Titration: A Method for Stability Investigations on MIEC Materials", oral presentation, *10th International Symposium on Ceramic Materials and Components for Energy and Environmental Applications (CMCEE)*, Dresden/Germany, May 2012.
- C. Niedrig, P. Braun, W. Menesklou, S. F. Wagner, and E. Ivers-Tiffée, "Investigation of MIEC Materials using an Amperometric Oxygen Titration Method", oral presentation, *221st Meeting of The Electrochemical Society (ECS)*, Seattle, WA/USA, May 2012.
- C. Niedrig, W. Menesklou, S. F. Wagner, W. A. Meulenberg, S. Baumann, and E. Ivers-Tiffée, "Influence of Oxygen Partial Pressure on the Conductivity and Stability of Selected MIEC Perovskites", oral presentation, *5th International Conference on Electroceramics (ICE)*, Sydney/Australia, December 2011.
- C. Niedrig, W. Menesklou, A. Weber, S. F. Wagner, and E. Ivers-Tiffée, "Electrochemical properties and stability of MIEC membrane materials", invited oral presentation, *12. Jülicher Werkstoffsymposium (WSP 12): "Gas Separation Membranes for Zero-emission Fossil Power Plants"*, Jülich/Germany, October 2011.
- P. Müller, H. Störmer, L. Dieterle, C. Niedrig, S. F. Wagner, D. Gerthsen, and E. Ivers-Tiffée, "Determination of the Phase Composition in $Ba_{0.5}Sr_{0.5}Co_{0.8}Fe_{0.2}O_{3-\delta}$ by means of Electron Microscopy", oral presentation, *18th International Conference on Solid State Ionics (SSI-18)*, Warsaw/Poland, July 2011.
- C. Niedrig, W. Menesklou, S. F. Wagner, S. Baumann, W. A. Meulenberg, and E. Ivers-Tiffée, "Electrical Conductivity and Stability of Selected Perovskite MIEC Materials at Low Oxygen Partial Pressures", oral presentation, *18th International Conference on Solid State Ionics (SSI-18)*, Warsaw/Poland, July 2011.

- C. Niedrig, W. Menesklou, S. F. Wagner, S. Baumann, W. A. Meulenberg, and E. Ivers-Tiffée, "$A_xSr_{1-x}Co_yFe_{1-y}O_{3-\delta}$ (A = La, Ba, Pr) for oxygen-transport membrane applications: Temperature and pO_2-dependent conductivity", poster presentation, *International Congress on Membranes and Membrane Processes (ICOM)*, Amsterdam, Netherlands, July 2011.
- C. Niedrig, S. Schütt, W. Menesklou, S. F. Wagner, S. Baumann, W. A. Meulenberg, and E. Ivers-Tiffée, "Stability of $A_xSr_{1-x}Co_yFe_{1-y}O_{3-\delta}$ (A = La, Ba, Pr) for oxygen-transport membrane applications", oral presentation, *12th Conference of the European Ceramic Society*, Stockholm/Sweden, June 2011.
- C. Niedrig, S. Taufall, S. F. Wagner, W. Menesklou, and E. Ivers-Tiffée, "Investigation of thermal and pO_2 stability of $Ba_{0.5}Sr_{0.5}Co_{0.8}Fe_{0.2}O_{3-\delta}$ for oxygen-transport membranes", poster presentation, *2nd International Conference on Energy Process Engineering (ICEPE)*, Frankfurt (Main)/Germany, June 2011.
- C. Niedrig, S. F. Wagner, W. Menesklou, W. A. Meulenberg, S. Baumann, and E. Ivers-Tiffée, "Conductivity and stability at low pO_2 of $A_xSr_{1-x}Co_yFe_{1-y}O_{3-\delta}$ (A = La, Ba, Pr) for gas separation membrane applications", oral presentation, *E-MRS Spring Meeting 2011*, Nice/France, May 2011.
- C. Niedrig, S. Taufall, H. Götz, S. F. Wagner, W. Menesklou, S. Baumann, and E. Ivers-Tiffée, "pO_2 stability of $Ba_{0.5}Sr_{0.5}Co_{0.8}Fe_{0.2}O_{3-\delta}$", poster presentation, *MRS Fall Meeting 2010*, Symposium EE, Boston MA/USA, November/December 2010.
- S. Taufall, C. Niedrig, H. Götz, S. F. Wagner, W. Menesklou, and E. Ivers-Tiffée, "Stability study of mixed-conducting $Ba_{0.5}Sr_{0.5}Co_{0.8}Fe_{0.2}O_{3-\delta}$ for oxygen-membrane applications", poster presentation, *MEM-BRAIN Summer School 2010: "Ceramic Membranes for Green Chemical Production and Clean Power Generation"*, Valencia/Spain, September 2010.
- S. Taufall, C. Niedrig, H. Götz, S. F. Wagner, W. Menesklou, and E. Ivers-Tiffée, "Stability study of mixed-conducting $Ba_{0.5}Sr_{0.5}Co_{0.8}Fe_{0.2}O_{3-\delta}$ for oxygen-membrane applications", poster presentation, *7th International Conference on Inorganic Materials*, Biarritz/France, September 2010.
- C. Niedrig, S. Taufall, S. F. Wagner, P. Müller, H. Störmer, D. Gerthsen, and E. Ivers-Tiffée, "Thermal Stability of mixed ionic-electronic conducting $Ba_{0.5}Sr_{0.5}Co_{0.8}Fe_{0.2}O_{3-\delta}$", oral presentation, *1st International Conference on Materials for Energy*, Karlsruhe/Germany, July 2010.
- C. Niedrig, S. Taufall, S. F. Wagner, E. Ivers-Tiffée, "Stability range of mixed ionic-electronic conducting $Ba_{0.5}Sr_{0.5}Co_{0.8}Fe_{0.2}O_{3-\delta}$", oral presentation, *Electroceramics XII*, Trondheim/Norway, June 2010.

- C. Niedrig, S. Roitsch, M. Burriel, W. Menesklou, S. F. Wagner, J. Santiso, J. Mayer, and E. Ivers-Tiffée, "Oxygen Exchange Kinetics and Structural Stability of Epitaxial $Ba_{0.5}Sr_{0.5}Co_{0.8}Fe_{0.2}O_{3-\delta}$ Thin Films", poster presentation, *E-MRS Spring Meeting 2010*, Strasbourg/France, June 2010.
- C. Niedrig, S. Taufall, S. F. Wagner, and E. Ivers-Tiffée, "$Ba_{0.5}Sr_{0.5}Co_{0.8}Fe_{0.2}O_{3-\delta}$ for oxygen separation membranes: A study of its thermal and pO_2-dependent stability", oral presentation, *Innovation for Sustainable Production 2010 (i-SUP2010)*, Bruges/Belgium, April 2010.
- P. Müller, L. Dieterle, E. Müller, H. Störmer, D. Gerthsen, C. Niedrig, S. Taufall, S. F. Wagner, and E. Ivers-Tiffée, "$Ba_{0.5}Sr_{0.5}Co_{0.8}Fe_{0.2}O_{3-\delta}$ for Oxygen Separation Membranes", oral presentation, *217th Meeting of The Electrochemical Society (ECS)*, Vancouver/Canada, April 2010.
- C. Niedrig, M. Burriel, S. Taufall, S. F. Wagner, P. Müller, H. Störmer, D. Gerthsen, and E. Ivers-Tiffée, "Transport Properties and Stability Issues of $Ba_{0.5}Sr_{0.5}Co_{0.8}Fe_{0.2}O_{3-\delta}$ (BSCF)", oral presentation, *4th International Conference on Electroceramics (ICE-2009)*, New Delhi/India, December 2009.
- M. Burriel, C. Niedrig, S. Baumann, S. F. Wagner, and E. Ivers-Tiffée, "Electrical and Oxygen Transport Properties of $Ba_{0.5}Sr_{0.5}Co_{0.8}Fe_{0.2}O_{3-\delta}$", oral presentation, *E-MRS Fall Meeting 2009*, Warsaw/Poland, September 2009.
- P. Müller, H. Störmer, D. Gerthsen, J. Santiso, J. Roqueta, M. Burriel, C. Niedrig, S. F. Wagner, and E. Ivers-Tiffée, "TEM investigations of mixed ionic/electronic conducting $Ba_{0.5}Sr_{0.5}Co_{0.8}Fe_{0.2}O_{3-\delta}$ materials", poster presentation, *Microscopy Conference 2009*, Graz/Austria, September 2009.
- P. Müller, H. Störmer, M. Burriel, C. Niedrig, S. F. Wagner, E. Ivers-Tiffée, and D. Gerthsen, "Microstructural and Electrochemical Properties of Mixed Ionic/Electronic Conducting (MIEC) $Ba_{0.5}Sr_{0.5}Co_{0.8}Fe_{0.2}O_{3-\delta}$ (BSCF) Perovskite Materials", poster presentation, *11th International Fischer Symposium on "Microscopy in Electrochemistry"*, Benediktbeuern/Germany, July 2009.
- C. Niedrig, M. Burriel López, S. F. Wagner, W. Menesklou, and E. Ivers-Tiffée, "Electrochemical Behaviour of $Ba_{0.5}Sr_{0.5}Co_{0.8}Fe_{0.2}O_{3-\delta}$ Thin Films", oral presentation, *17th International Conference on Solid State Ionics (SSI-17)*, Toronto/Canada, June/July 2009.
- M. Burriel, C. Niedrig, S. F. Wagner, W. Menesklou, and E. Ivers-Tiffée, "A study of the electrochemical properties of $Ba_{0.5}Sr_{0.5}Co_{0.8}Fe_{0.2}O_{3-\delta}$", oral presentation, *MRS Fall Meeting 2008*, Boston MA/USA, December 2008.

B.3 Supervised Student Theses

- Lennart Willms, "Numerische Analyse der elektrochemischen Übertragungsfunktion von Mischleitern", *Bachelor Thesis* (in German), Institut für Werkstoffe der Elektrotechnik (IWE), Karlsruhe Institute of Technology (KIT), August 2013.
- Lana-Simone Unger, "Elektrische Messungen an kationisch teilsubstituierten Hochtemperatur-Mischleitern auf Basis von BSCF5582", *Study Thesis* (in German), Institut für Werkstoffe der Elektrotechnik (IWE), Karlsruhe Institute of Technology (KIT), July 2012.
- Isabelle Tanja Hofstetter, "Methodische Messwert-Analyse der amperometrischen Sauerstofftitration", *Bachelor Thesis* (in German), Institut für Werkstoffe der Elektrotechnik (IWE), Karlsruhe Institute of Technology (KIT), January 2012.
- Philipp Braun, "Einsatz der amperometrischen Sauerstofftitration zur Charakterisierung von HT-MIEC", *Bachelor Thesis* (in German), Institut für Werkstoffe der Elektrotechnik (IWE), Karlsruhe Institute of Technology (KIT), November 2011.
- Simon Taufall, "Untersuchung der elektrochemischen Stabilität von BSCF-Sauerstoffseparationsmembranen", *Diploma Thesis* (in German), Institut für Werkstoffe der Elektrotechnik (IWE), Karlsruhe Institute of Technology (KIT), April 2010.

Werkstoffwissenschaft @ Elektrotechnik /
Universität Karlsruhe, Institut für Werkstoffe der Elektrotechnik

Die Bände sind im Verlagshaus Mainz (Aachen) erschienen.

Band 1 Helge Schichlein
Experimentelle Modellbildung für die Hochtemperatur-Brennstoffzelle SOFC. 2003
ISBN 3-86130-229-2

Band 2 Dirk Herbstritt
Entwicklung und Optimierung einer leistungsfähigen Kathodenstruktur für die Hochtemperatur-Brennstoffzelle SOFC. 2003
ISBN 3-86130-230-6

Band 3 Frédéric Zimmermann
Steuerbare Mikrowellendielektrika aus ferroelektrischen Dickschichten. 2003
ISBN 3-86130-231-4

Band 4 Barbara Hippauf
Kinetik von selbsttragenden, offenporösen Sauerstoffsensoren auf der Basis von $Sr(Ti,Fe)O_3$. 2005
ISBN 3-86130-232-2

Band 5 Daniel Fouquet
Einsatz von Kohlenwasserstoffen in der Hochtemperatur-Brennstoffzelle SOFC. 2005
ISBN 3-86130-233-0

Band 6 Volker Fischer
Nanoskalige Nioboxidschichten für den Einsatz in hochkapazitiven Niob-Elektrolytkondensatoren. 2005
ISBN 3-86130-234-9

Band 7 Thomas Schneider
Strontiumtitanferrit-Abgassensoren. Stabilitätsgrenzen / Betriebsfelder. 2005
ISBN 3-86130-235-7

Band 8 Markus J. Heneka
Alterung der Festelektrolyt-Brennstoffzelle unter thermischen und elektrischen Lastwechseln. 2006
ISBN 3-86130-236-5

Band 9 Thilo Hilpert
Elektrische Charakterisierung von Wärmedämmschichten mittels Impedanzspektroskopie. 2007
ISBN 3-86130-237-3

Band 10	Michael Becker **Parameterstudie zur Langzeitbeständigkeit von Hochtemperaturbrennstoffzellen (SOFC).** 2007 ISBN 3-86130-239-X
Band 11	Jin Xu **Nonlinear Dielectric Thin Films for Tunable Microwave Applications.** 2007 ISBN 3-86130-238-1
Band 12	Patrick König **Modellgestützte Analyse und Simulation von stationären Brennstoffzellensystemen.** 2007 ISBN 3-86130-241-1
Band 13	Steffen Eccarius **Approaches to Passive Operation of a Direct Methanol Fuel Cell.** 2007 ISBN 3-86130-242-X

Fortführung als

Schriften des Instituts für Werkstoffe der Elektrotechnik, Karlsruher Institut für Technologie (ISSN 1868-1603)

bei KIT Scientific Publishing

Die Bände sind unter www.ksp.kit.edu als PDF frei verfügbar oder als Druckausgabe bestellbar.

Band 14	Stefan F. Wagner **Untersuchungen zur Kinetik des Sauerstoffaustauschs an modifizierten Perowskitgrenzflächen.** 2009 ISBN 978-3-86644-362-4
Band 15	Christoph Peters **Grain-Size Effects in Nanoscaled Electrolyte and Cathode Thin Films for Solid Oxide Fuel Cells (SOFC).** 2009 ISBN 978-3-86644-336-5
Band 16	Bernd Rüger **Mikrostrukturmodellierung von Elektroden für die Festelektrolytbrennstoffzelle.** 2009 ISBN 978-3-86644-409-6
Band 17	Henrik Timmermann **Untersuchungen zum Einsatz von Reformat aus flüssigen Kohlenwasserstoffen in der Hochtemperaturbrennstoffzelle SOFC.** 2010 ISBN 978-3-86644-478-2

Band 18 André Leonide
SOFC Modelling and Parameter Identification by Means of Impedance Spectroscopy. 2010
ISBN 978-3-86644-538-3

Band 19 Cornelia Endler-Schuck
Alterungsverhalten mischleitender LSCF Kathoden für Hochtemperatur-Festoxid-Brennstoffzellen (SOFCs). 2011
ISBN 978-3-86644-652-6

Band 20 Annika Utz
The Electrochemical Oxidation of H_2 and CO at Patterned Ni Anodes of SOFCs. 2011
ISBN 978-3-86644-686-1

Band 21 Jan Hayd
Nanoskalige Kathoden für den Einsatz in Festelektrolyt-Brennstoffzellen bei abgesenkten Betriebstemperaturen. 2012
ISBN 978-3-86644-838-4

Band 22 Michael Kornely
Elektrische Charakterisierung und Modellierung von metallischen Interkonnektoren (MIC) des SOFC-Stacks. 2012
ISBN 978-3-86644-833-9

Band 23 Dino Klotz
Characterization and Modeling of Electrochemical Energy Conversion Systems by Impedance Techniques. 2012
ISBN 978-3-86644-903-9

Band 24 Alexander Kromp
Model-based Interpretation of the Performance and Degradation of Reformate Fueled Solid Oxide Fuel Cells. 2013
ISBN 978-3-7315-0006-3

Band 25 Jan Philipp Schmidt
Verfahren zur Charakterisierung und Modellierung von Lithium-Ionen Zellen. 2013
ISBN 978-3-7315-0115-2

Band 26 Moses Ender
Mikrostrukturelle Charakterisierung, Modellentwicklung und Simulation poröser Elektroden für Lithiumionenzellen. 2014
ISBN 978-3-7315-0205-0

Band 27 Jörg Illig
Physically based Impedance Modelling of Lithium-Ion Cells. 2014
ISBN 978-3-7315-0246-3

Fortführung als

Schriften des Instituts für Angewandte Materialien – Werkstoffe der Elektrotechnik Karlsruher Institut für Technologie (ISSN 2365-8029)

bei KIT Scientific Publishing

Die Bände sind unter www.ksp.kit.edu als PDF frei verfügbar oder als Druckausgabe bestellbar.

Band 28 Christian Niedrig
Electrochemical Performance and Stability of $Ba_{0.5}Sr_{0.5}Co_{0.8}Fe_{0.2}O_{3-\delta}$ for Oxygen Transport Membranes. 2015
ISBN 978-3-7315-0437-5